APPLIED FIELD SURVEYING

APPLIED FIELD SURVEYING

KING ROYER

John Wiley & Sons, Inc.
New York · London · Sydney · Toronto

Library of Congress Catalogue Card Number: 70-106014

SBN 471 74395 X

Printed in the United States of America

10 9 8 7 6 5 4 3 2 1

PREFACE

The completion of a construction project requires three kinds of surveys:

1. *A survey to find out where the property is.* This is called a boundary, land, or cadastral survey, and the work is done by a registered land surveyor.
2. *A survey to find out what the existing conditions are.* This may be mapping (by a preliminary survey of any kind) or merely an inspection of the property. For highways, pipelines, and canals, it is a map of a long narrow strip on which the work is located. If this survey is made by a separate firm, it will be by a registered professional engineer, although a land surveyor may do this kind of work in connection with boundary surveys.
3. *A survey to show the location and elevation of construction work.* This survey may be elaborate; for example, for highways, where survey crews work on such projects for years. For a house, this survey may be made by a carpenter using a carpenter's tape, string, and level.

Sometimes these surveys are made by separate crews; at other times one crew will make all three types. The methods and tools that are used are the same in any case, although different training is required of people who direct the work. This book is for those who will do ordinary work common to all types of surveying, under the supervision of an experienced chief of a survey party or a job superintendent.

For the last few decades, most surveying courses have been offered as part of the engineering curricula. In these courses and in the textbooks written for them, computations and precise work are emphasized. This book is a guide for students who want to learn the minimum theory necessary to be useful on a field job. For those who are to advance to supervisory positions, this course is an introduction to further study either in school or on the job. For carpenters, superintendents, operating engineers, and other building tradesmen who use these skills in comparatively simple situations and as a minor part of their daily work, the explanations in this book are adequate. For highway engineers and land surveyors, it is an introduction to practical procedures.

The book may be used in a 14-week course of two 4-hour field periods per week, with little homework required. Many of the exercises may be completed in less than 4 hours, and some may be made shorter or longer to meet the time available. The book may be readily adapted to a continuous 2-week course of 40 hours per week.

Field exercises 1 to 6 give basic practice in instrument use. They may be rearranged so that part of the class is practicing chaining, part transit work, and part level work, at the same time. In this way, individual instruction can be provided during the early part of the course, and less supervision is necessary in the latter portion. At the University of Florida, each student is given a transit or level for the first exercises. Close supervision of students is necessary during this period. If a series of courses is to be held, the courses may overlap, making best use of the instructors and the equipment.

The Classroom Exercises may be performed when weather conditions are suitable. The traverse and curve calculations (Numbers 4 and 6) may be made at any time, and other Classroom Exercises may, in general, be completed as soon as a few field periods have provided data. At Florida, the first 13 Field Exercises are completed together, weather permitting, so that there can be ample work remaining for rainy days. The stadia and pipeline surveys (Numbers 7 and 8) take two periods each; Numbers 14, 15, and 16 can sometimes all be completed in a single period.

For a two-quarter course with one field period per week, the greater part of the classroom exercises can be completed as homework. This course may be taught as a two-week course by omitting some work or by working on Saturdays. In a short course, work done in the field is more efficient because less time is lost in traveling, checking equipment, and getting started with each exercise.

Graduates of this course should be able to work as instrument men or rodmen on a survey party, under the supervision of a party chief, and advance to party chief as ability permits. For use in an engineering-degree program, this book is designed as a laboratory manual, to be used with a standard engineering surveying textbook, and with theoretical work assigned as homework or taught in a separate course.

Emphasis is placed here on individual work as much as possible, and on work that may be checked by the student himself. The first exercises are of this type. The construction aspects of surveying are also stressed; a recent textbook author commented that 60 percent of all surveying manhours were spent on giving line and grade, but in his own book devoted

98 percent of the space to other topics! "Engineering surveying" has been construed to mean mapping and a certain amount of route surveying; the railway engineering boom of the late 1800's has not exhausted its influence. Mapping emphasis requires very little of the instructor, but it also gives the student very little training. The traditional stadia map has been retained in this textbook, but to a reduced degree, and with emphasis on the ability of the individual to make a stadia shot—often not acquired by the bottom man in a five-man party. The traverse has been abandoned altogether as wasteful of time for the amount of work done, and as not adapted to checking the work of one person rather than of the party.

I suggest that the better students and men with surveying experience should be selected in the first class and should be assigned to work together. The practice of using experienced students as field party leaders often results in their monopolizing the work to the detriment of the less accomplished men. In this way the party makes a good showing, but many students practice the work very little. If the better men are placed together, they will be more patient with the repetition that occurs for them and will, to some extent, set standards for the rest of the class.

It is recognized that there are many ways of accomplishing the work shown here, and criticism and comments to help improve future editions are invited. There are numerous statements in the book that oversimplify complicated matters or that give the impression that the stated method is the only method or even the accepted one. This is done because this book is not intended as an academic contribution, but as a simplification of methods that have been stated countless times before. My intention is to give information to first-semester students without needlessly confusing them.

The sources from which items in this book have been taken are too numerous to mention, but I am especially grateful to Joe Butler, Highway Location Engineer of the State Road Department of Florida, under whom I served as rodman, instrument man, and chief of party, and Professor H. H. Block of the University of Florida, under whom this book was prepared as a series of course notes for surveying classes over a period of six years.

Gainesville, Florida, 1970 KING ROYER

CONTENTS

APPLIED FIELD SURVEYING

Chapter I

TAPES AND CHAINS

Before inexpensive steel tapes were produced, surveyors used a steel chain of 100 links for measurement. Although steel tapes are now used almost exclusively, the term *chain* or *chain tape* still applies to the heavy-duty steel tape used by engineers and surveyors. It is also sometimes called an *engineer's tape*. *Chaining,* therefore, means the process of measurement by using the chain. The words chaining and taping are used interchangeably.

The old surveyor's measure was the *Gunter's chain,* an English measure of 66 ft. This measure is the basis of modern land distance and area units. An area one chain square is 0.10 acre. A *rod* is a fourth of a chain; fencing is sold by rods. There are 80 chains to a mile, and 10 chains to a *furlong.* A furlong, or city block in many areas, was the length of furrow that could be plowed by an ox without resting. Therefore, a field which could be efficiently plowed each way was a furlong, or 10 chains, on a side, and contained 10 acres.

The term *chain* is occasionally (and loosely) used in modern surveying to indicate a hundred feet. More properly, it indicates the old 66-ft Gunter's chain used in old land descriptions. The measurements in these old descriptions must frequently be recomputed for land surveying. Since these units were used in the original surveys, they are in permanent land records and may not be changed.

Three kinds of measuring tapes are in common use: the chain or chain tape, the carpenter's tape, and the metallic tape.

THE CHAIN

The chain is most useful in measuring long distances, where there is no support for the tape (as when measuring across a hole), and in muddy and sandy locations. It may be looped and unlooped, similar to the way a cowboy loops his lariat, more rapidly than wound up on a reel; it is used both with and without the reel. It may be looped or wound up without removing mud and sand, unlike other tapes which are wound into a closed case.

The chain is very difficult to break under direct pull. It can be easily spliced on the job if broken, and has permanent graduations stamped into the steel itself or into lead (or more properly, babbitt) pieces on the steel. It may be dragged from one place to another without damage, even behind a truck if the surface of the road is not too abrasive. Because of this, it is sometimes called a *drag tape*.

The chain has several disadvantages when used for construction work.

1. For small jobs, a carpenter's tape is more easily stored in a tool box, and is cheaper. A 100-ft chain can cost nearly twice as much as a carpenter's tape of the same length, and the cost of the carpenter's tape includes the case. The carpenter's tape is available at hardware stores, although chains are sold only at engineering supply houses.
2. Apprenticeship training does not include use of the chain, so most superintendents and carpenter foremen are not accustomed to it.
3. The chain is dimensioned only on the first and last foot, rather than throughout its length as is a carpenter's tape, and is marked in hundredths of a foot. For short measurements, hundredths must be converted to inches for building work.
4. Unless the chain is on a reel, it is difficult to measure distances less than 100 ft, because the entire length must be let out.
5. The chain is stiff, and it is easily kinked and ruined by being doubled back on itself.

For these reasons, the chain is usually used only on larger projects which require a surveyor, or where special accuracy is necessary.

The chain will take any kind of rough wear except kinking, to which it is very sensitive. It may be yanked out of the way of approaching traffic and allowed to land in a pile, but it must then be carefully untangled. It may be pulled around an arm or leg to change direction, but not around smaller objects. Twists are harmful, because they may turn into loops and be pulled into kinks. The chain is undamaged by salt or fresh water, but when used in salt water should be cleaned at the end of the day.

Chains have graduations at the first and last foot, either pressed into lead, in which case markings are to tenths of feet only, or pressed into the steel of the tape. Steel markings on the first and last foot are to hundredths of a foot, but these corrode easily and must be cleaned by sanding periodically. Chain tapes with graduations throughout their length are available, and in 1965 about 10% of all chains sold were of this type.

A chain may be spliced by an inexpensive steel splice sold for this purpose. The ends are flattened, placed together on a hammer or convenient rock, and the splice is clinched over the ends with a hammer. The splice is then heated with a match to melt the solder in the joint and to make a strong connection. Kinks are never entirely eliminated. The chain may be bent out, so kinks are not apparent when the chain is extended, but they will cause loops to form in the wrong direction when the chain is looped up. Chains usually accumulate kinks and splices until they become difficult to use and are then junked.

When the chain is wound on a reel, a leather thong can be tied to only one end of the chain; the other thong is left on the handle of the reel. The end of the chain in the reel fits a projection on the reel. When the chain is used as a drag tape, thongs are tied to each end; these leather thongs are furnished with the new tape. The thong at the leading end, ordinarily the zero end, has a slip knot to fasten around the lead chainman's wrist, so he need not pick up the end from the ground. The thong on the following, or tail end, is free in order to avoid looping on roots and bushes.

Figure 1*a*. Looping the chain the right way.

Figure 1*b*. Looping the chain the wrong way.

When the reel is not used, the chain is tied in 5-ft loops. When let out from the loop, it may be let out from either end. Picking up, looping, and unlooping a chain, although not complicated, must be taught by an instructor or learned from movies. Remember that a chain must be done up "right-handed" (Figures 1 and 2), even for left-handed persons, if more than one chainman is to use the same chain. If the chain is looped right-handed and unlooped left-handed, a tangle immediately results. The chainman should learn to test the chain by checking the first loop released to be sure it is in the right direction. An experienced chainman will recognize a chain put up left-handed almost as soon as he picks it up.

THE CARPENTER'S TAPE

There are a variety of lengths and colors of carpenter's steel tapes; cheap ones have painted numbers, and others have more durable markings. Usually they have a hook on the end, so that one person may make measurements. They must not be dragged, and must be cleaned carefully before being wound in the case. They may be carried in a plastic carrier

Figure 2. Snapping the chain.

on the belt; tapes varying from 25 to 100 ft are available in the same size case.

THE METALLIC TAPE

The metallic tape is of canvas, with interwoven metal threads to prevent stretching. It is sold in a case similar to that for a carpenter's tape. The usual length is 50 ft, graduated in feet and tenths throughout its length. This tape is extensively used for approximate measurements, such as topographical mapping, when measurements to trees, buildings, and similar landmarks are made, and in measuring to the level rod when elevations of the ground are taken. The markings wear off eventually, and refills for the case are available. This tape is particularly useful for many short measurements, as it may be looped like a lariat and allowed to pass out through the fingers. Such treatment of a steel tape would ruin it in a very short time.

FIELD NOTES

The notes kept for land surveys and building construction layout are drawings that show measurements and angles. Notes for many kinds of engineering mapping and construction have been reduced to routine, and are standardized by use of certain accepted symbols and methods of recording data, so an actual picture is not necessary. Notes for pipelines, highways, and ditches are of this type. Such standardized notes are simple and easily understood by a draftsman. Some of these notes are shown in this manual, particularly those for stadia surveying (mapping) and for pipeline surveys, but no attempt is made at completeness. When you start work on a field crew, you will have to learn its work and exactly how that crew takes its notes.

All types of field notes have certain points in common. The purpose of the notes is to show, to a person with ordinary engineering knowledge, what work has been done. No surveyor's notes should be so vague that he has to explain what he means by a telephone call or a trip to the office. Some of the common features of notekeeping are listed here.

1. All notations made in the field, if they are written down at all, should be in the notes, whether important or not. "Scratch sheets" are for beginners, not for engineers or surveyors. If no space is available on the note pages in use, computations are made on the back pages of the book. Notes should not be recopied, unless the duplicates are marked *copies,* and the original notes should always be preserved.

2. All measurements of angles and distances should be written down as soon as they are made. No additions or subtractions should be made before the figures are entered in the book; for example, if the curb of a street is 15 ft from the centerline and you are measuring from the curb, show the actual measurements from the curb that you read on the tape; do not add 15 ft and then record the reading. If you do, there is no way to check later and correct an error. You can hold the 15-ft mark on the curb, and read a distance on the tape which represents the distance to the centerline. Remember it is not necessarily the actual distance you are recording; it is the *reading* that you get.

3. When you make a drawing beforehand showing what dimensions *should* be, as for layout of a house, do not let these dimensions become confused with the *actual* measurements you make. They should be on different pages, or one set should be marked *computed* and the other set *actual.* Do not fail to write down a measurement because it is what it is supposed to be; the draftsman cannot tell if you checked it, and later on you might not remember it yourself.

4. Record each page number in the upper right-hand of each right page. Unlike a page of a book, a page in a surveying field book includes both the right- and left-hand sides exposed at one time. The names of the survey crew and date are shown at the beginning of each day, when a new

book is started, when a new kind of work is begun, or when the notes skip from one part of a book to another. This enables the draftsman to find out who was doing each particular job, which may save repeating a field measurement to locate an error. The first part of each book is the index page, and here you should list all notes in the book.

STUDENT FIELD NOTES

If the exercises in this book are assigned, your instructor will give you specific instructions. If men work in crews, for example, it may be desirable that notes be copied by other members of the party for practice purposes. Notes should not be kept in any form other than in the field book; a loose-leaf book is suggested, so that the page may be handed in and returned the next period while you still have the book. Notes should be completed at the end of each period, and clip boards or other kinds of notebooks should not be used by students. Students have a tendency to make notes "pretty" which makes the notes inaccurate and leads the students to do sloppy work in the field. You will not be able to take useful notes on a survey crew unless you learn to write clear figures under field conditions.

The instrument number should also be recorded in your notes. If an instrument is out of adjustment, it can later be checked if you cannot account for errors in your work in any other way. If the instructor finds that an instrument is improperly stored in the box, he can tell from the field notes who is responsible. When doing level work in a group where different kinds of instruments are in use, such as builder's levels and engineer's levels, the kind of level should be shown in the notes to justify the degree of accuracy attained in the work.

FIELD EXERCISE NUMBER 1

It is anticipated that the first field period will be taken up largely by the instructor's lecture, so this exercise may be completed in as little as thirty minutes. If field books are not obtained before the first field period, it may be necessary to take notes on notebook paper for this exercise only.

This is individual work; each student requires a chain tape, with end thongs.

Let out a chain and pick it up again as instructed in class. Record the time you first start and the time at which you start each subsequent repetition, remembering that the actual watch reading, not an imaginary reading such as zero, is to be recorded. Alternately let out the chain from the 0

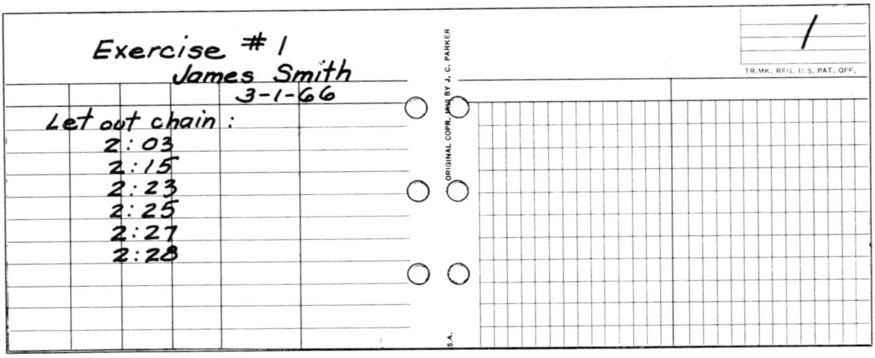

Figure 3. Notes for Exercise Number 1.

and from the 100-ft end by turning it over in your hand. Continue this operation until you can let out a 100-ft chain and pick it up, completely bound, twice in ten minutes. The time is actually not important; if you can do it at all, you can do it in less than five minutes. The better students may have a race with each other at this operation, and will soon be able to complete the operation faster than they can walk the length of the chain. Do the following for each cycle.

1. Turn the chain over to reverse ends.

2. To let out, hold in left hand, point loose end toward yourself, and let out to right.

3. When picking up, be sure to start at the 0 or 100-ft mark. Some chains are 101 ft long, and the −1-ft mark is easily mistaken for the zero. Hold the chain in exactly the same position when picking it up as when you let it out. It does not matter what direction you face when picking up the chain, or whether you stand in one place or walk in any direction. Watch your own hands; the chain will follow the way you pull it.

4. When picking up, lay each 5-ft mark on top of the next. You can reach further than 5 ft unless you are quite short. You should learn to reach about 5½ ft each time, which will put the mark where you want it.

5. When lashing the ends, be sure that each end is wrapped *back* on itself, and pull the lashing tight. A loosely lashed chain will come apart later and become tangled.

Some common student errors are listed here.

1. Rather than pick up the chain from right to left, a student will hold the looped portion across his body and pull the strands toward himself as

though he were rowing a boat. This is slow and laborious, and often results in the chain being looped left-handed. It is quite possible to hold the chain backwards and then do it up backwards, so the result may be correct. A strongly left-handed person may be obliged to work this way: he would use the method marked "wrong way" in Figure 1, but with *hands reversed to the illustration*.

2. Sometimes a student will lay several turns of the chain down on the ground before looking to see if it lies flat. This can result in a scrambled tape. If you have loops in the tape when you start, don't continue, but ask for help. You can't reverse an operation once started, and it may take three hands to get you straightened out.

3. A man often attempts to align his body to the chain already on the ground. This doesn't matter, except that his attention will be on the ground, rather than on his hands, where the action is. The tape on the ground can be forgotten (unless you attempt to let it all out at one place, and then step on it!).

4. A student who wants to start from the zero end, for example, will look inside the loop, see the end he wants, and try to let it out from the inside. This can be done, but amateurs should not do it, and it is unnecessary. Turn the tape by turning the wrist over, holding the tape firmly; then the desired mark will be on top.

Chapter 2

CHAINING

The method of measuring longer distances described in this chapter provides reasonable accuracy with the greatest practical speed. Speed in a manual operation is attained primarily by reducing the number of motions to as few as possible, and then following these motions in rhythm. In surveying, speed is attained first of all by avoiding unnecessary repetition. It is not merely sufficient to be able to measure a distance; even a beginning chainman should be able to chain a 1000-ft line in fifteen minutes, if there are no slopes requiring extra operations.

SHORT MEASUREMENTS

There are two ways to read a chain for distances under 100 ft, depending on whether a 100- or 101-ft chain is used. For example, if two points 65.68 feet apart are being measured with the 100-ft chain, the 66-ft mark must be placed on one point; the other rodman reads the distance of his point from the end of the chain, which is 0.32 ft. The first rodman calls out "66 feet," the second "cutting thirty-two"; a rodman or notekeeper then calls out "65.68" so the other may check his subtraction, and this figure is entered in the notes. This procedure is an exception to the rule that only actual numbers read are recorded in the notes.

If a 101-ft tape is used, one rodman holds at 65 ft, and the second reads 0.68 feet on his additional foot. Each calls his own reading, and this is entered in the notes as read. The 101-ft tape is, therefore, easier to use and more accurate; but in chaining long distances the extra foot is in the way, and a chainman may drive his stake at the −1-ft mark rather than at zero. This is a rather common error, especially for beginners; the reported distance will then be 1 ft too small for each full 100 ft measured.

LINE AND LEVEL

Horizontal lines are measured in a straight line. In accurate work and where stakes are being set for later use, as on a highway centerline, points may be lined up with a transit to any desired accuracy. For determining the length of a line, as in boundary surveying, the points must be set close enough to a straight line that the error in length is not significant. Surveying of the type described in this book is seldom accurate to over 1/10,000, or 0.01 ft (practically 1/8 in.) in a 100-ft chain. This accuracy can be attained when a stake is as much as 1.4 ft off the line of the previous stake. It is undesirable to make any avoidable error, but if the stakes are within 0.5 ft of the true line, error is only about 0.001 ft—an insignificant amount. For accuracy in measuring, therefore, alinement by eye is sufficient.

Lines measured to a point off line, as around a tree, may be adjusted as shown in Figure 5. In a 100-ft length, the error on the line will be 0.005 ft multiplied by the square of the distance off line at the end. If a stake is 3-ft off line, the error in 100 ft is 0.005 ft $\times$ 3 $\times$ 3 = 0.045 ft. For other distances than 100 ft,

$$\text{Error} = \frac{(\text{distance off line in feet})^2}{2 \times (\text{length of line in feet})}$$

This formula is not exactly correct, and becomes less accurate as the distance off line increases. It will give results within 0.01 ft for a distance off line as great as 12 ft in 100 ft.

Distance, unless otherwise stated, always refers to the distance as measured on a level line. If a line is not level, the measured distance will be greater than the actual horizontal distance, in accordance with the formula given above. Sometimes distances are measured on a slope, but field work and computations are usually easier if all lines are measured with the chain level. It is hard to judge a level line, unless there is another level line, such as a body of water, or a horizontal line on a building, to compare it with. Fortunately, a line may depart from the level as outlined above for all line points, so that if a line is within a foot of level the error is not substantial. This degree of accuracy may be achieved by trying several elevations each two feet from the preceding one, and using the one nearest level, as illustrated in Figure 4. In chaining, the point obtained which is *furthest* from the starting point, for a 100-ft chain, is the most nearly correct (that is, the point that appears furthest when it is level, plumbing from the ends, as in Figure 5). When making a measurement between two fixed points, the reading on the tape which gives the *least* distance between the points (again, when the *tape* is *longest*) is the correct measurement.

Using this rule, stakes may be driven on the slope by estimating the elevation difference. Since a stake is usually ⅝ in. wide, or 0.05 ft, the stake may be driven with the back face (nearest the tail chainman) to the

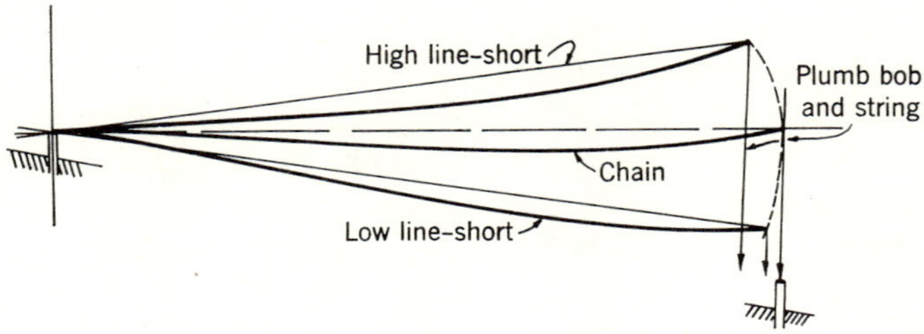

Figure 4. The chain will be level when plumb bob reaches furthest to the right.

slope distance, and the true distance, obtained by plumbing, will fall on the stake. The distance may then be marked with a plumb bob point, or in less accurate work, by calling "back," "middle," or "front" of stake. This avoids a preliminary operation of plumbing to determine where to drive the stake.

CORRECTIONS

Chains are usually true length at a temperature of 68 degrees with a tension of 10 lb when fully supported, as when lying on a pavement. If the temperature is higher, the chain will be longer by about 0.01 ft for each 15 degrees increase in temperature; and for colder weather, it will be shorter by a corresponding amount. For accurate work, this should be taken into account. If the chain is supported only at the ends, which is common during chaining, sag will bring the ends closer together and it will appear to measure less than 100 ft. If the chainmen pull harder on the ends, this sag is less, and in addition, the chain is stretched so that the true length along the curve is more than 100 ft. These effects compensate, so that there is an amount of sag where there is no correction for either sag or pull. This may be found by measuring two points on the ground 100 ft apart by supporting the chain, then raising the chain and measur-

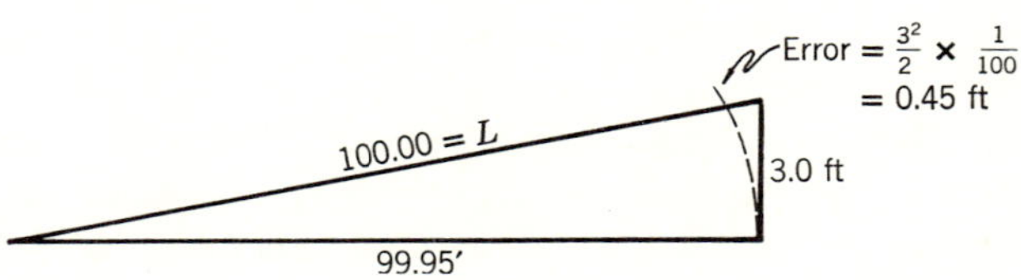

Figure 5. Error in measurement by stake off line.

ing the points again with the chain unsupported. It will be found that with about 18 in. to 24 in. sag in a 100-ft chain, the distance between the ends will be correct. This operation makes determination of pull unnecessary; and even if pull is measured, sag must still be known or computed. Although pull for an unsupported chain varies with the weight of the chain, correct sag as determined in this way is the same for any steel tape.

CHAINING A LONG LINE

When a long line is to be measured, the head chainman starts off with ten stakes, and the tail chainman has one. When 1000 ft have been measured the head chainman is out of stakes. The tail chainman, who has been removing used stakes, gives his stakes to the head chainman, who can then measure another 1000 ft (an extra stake is in the ground when the transfer is made). In this way only the number of thousands of feet need be recorded. In land surveying of sections, years ago, the stakes were brought forward every furlong (10 chains), and a notch was made in a stake to keep count; the half-mile post was set every four notches. Today, exact points are set with a nail in a larger stake (hub) every thousand feet on centerlines of highway and pipeline surveys.

OPERATIONS OF HEAD CHAINMAN

The head, or lead, chainman is the man on a survey crew who determines the accuracy of the work and sets the pace. He is therefore the best available man; many surveyors act as their own head chainman, leaving routine instrument work to a less-skilled man.

The equipment of the head chainman consists of these tools.

1. *The chain.* The zero end is attached to his left wrist (if he is right-handed) with a looped thong; thus he will not have to bend to pick up the end, and he can carry stakes in the same hand. With the zero end forward, measurements may be made by the head chainman stopping at the required distance. If the head chainman carries the 100-ft end of the chain, he must walk the full 100 ft, drop the chain, and walk back to his required foot. He must then return to the zero end, put it around his wrist, and continue. However, if a number of measurements are made within each 100 ft, this procedure may be faster.

2. *The hatchet,* if the stakes are being used. A carpenter's broad hatchet has a head and weight as great as that of a small axe, with a cutting blade wide enough to cut out roots and other obstructions to the stake. It has a sharp point on the edge of the cutting blade, which is helpful in cutting through ice or frozen ground. A camper's axe, or a short-handled axe, is

also sometimes used. The hatchet is carried in the right hand (for a right-handed man), and if this hand is used for plumbing, the hatchet is placed on the left side of the line being run, that is, on the right side of the chainman when he is facing the rear.

3. The *range rod,* if one is being used, is also carried in the right hand. If no points are to be set exactly on line, or if the distance is short enough that a plumb bob can be seen from the instrument (two or three hundred feet, unless a target on the string is used, or lighting conditions are particularly favorable) the rod is not required. The range rod is about 6-ft long, with alternating red and white stripes each exactly 1-ft long. It may be of wood or aluminum, and may be made in sections for disassembly and packing in a car.

4. The *plumb bob* is of brass with a removable steel tip to keep the point sharp when marking hard materials. It is on a piece of string as long as the chainman can reach, or longer. The best string is braided nylon mason's string, or nylon fish line. Bright orange string is available for better visibility. Twisted line will do, but comes untwisted with use. The bob is carried in a holster on the right hip, and during chaining the string is looped around the chainman's neck. It can then be lifted out of the holster with the right hand in a single motion, letting the string slide through the fingers. When plumbing from a mark on the chain, a lower finger touches the chain to steady the string and the string is brought to the mark. Do not lay the string on the chain except in the final stage of measurement, as vibrations in the chain are reflected in the plumb bob and make it swing widely.

5. The *stakes* are carried in the left hand, or cradled in the left arm. They are laid on the ground, *after* the chainman has his approximate line, has turned around, and knows where his stake will be, so the hand is free to work. The chainman then will not have to crawl through the mud on all fours, after driving one stake, to pick up the others.

6. A *machete* may be carried in the belt or in a scabbard, if it is likely to be needed. If it will be needed only rarely, it may be carried by the rear chainman.

The head chainman in a cycle of each 100 ft, in continuous chaining, performs the following cycle of operations.

1. He drags the chain, and paces off 100 ft.

2. When he is near 100 ft, he should hear the rear chainman call "chain." The head chainman turns, without dropping his equipment, and puts the chain as near on line as he can judge. If he is using a range rod, he will use the right hand to put it on line. He receives line correction by signal from the rear chainman, or on more accurate work, by the transitman.

3. He places his stakes, except the one to be driven, to his left. Keeping the chain tight with his left hand, he puts the stake on line and distance with the right hand, lining it up with the range rod in the ground or getting line correction from the rear chainman.

4. He drives the stake as nearly vertical as possible, starting measurement at the bottom of the stake, and shifting the chain to the top of the stake: without relaxing the pull on the chain, he stops driving the stake when it is firm in the ground, with the edge of the stake furthest from him on the exact mark on the chain.

The stake is hit only on the top, not on the side. If you must shift the stake in any direction, hit the ground on the opposite side, not the stake itself. Otherwise you will loosen the stake in the ground. Do not attempt to drive stakes from a standing position, as you cannot easily reach your tools, and you will drive the stake toward you. A squatting position, with one leg up and the other knee on the ground is most stable and least tiring, as in Figure 6a, not as in 6b. Get your body on line, so the left forearm is in line with the chain; do not try to pull the chain across your body or toward your stomach.

5. The chainman checks the stake, plumbing if necessary, as in Figure 7. He selects the edge or middle of the stake, calling out "your side," "my side," or "center" to the rear chainman to indicate the exact point

Figure 6*a*. Driving a stake the right way. The stake is driven directly downward, with the chain held on the mark with the left hand.

Figure 6*b*. Driving a stake the wrong way. The chainman, while standing, drives the stake toward himself, and has no mark to tell him when it is driven to correct distance.

of measurement. This is the signal to the rear chainman that he may release the chain and pull his own stake, if it is not a permanent marker.

6. The chainman picks up his stakes and rod, turns to the right (so he will not wind the chain around his body) and continues to the next stake.

• OPERATIONS OF REAR CHAINMAN

As the head chainman must be aggressive, the rear chainman must be patient. The rear chainman can avoid delaying the work, but cannot make it go faster.

The rear chainman carries a plumb bob, and the stakes that he pulls up as he goes. Since he has less to do than the head chainman, he carries any equipment which is needed—extra stakes, a machete, water, a file, a crayon, or perhaps lunches. In each 100-ft cycle he performs the following cycle of operations.

1. He stands at the rear stake as the chain passes by, picking it up as the end approaches him. When about 6 ft to the end remains, he calls "chain" to tell the head chainman he must stop after the next step. The rear chainman has one leg stiffly behind the stake, the other just off line as in Figure 8, so that he will not be pulled beyond his point or yield to jerks of the chain.

Figure 7. The lead chainman holds the chain in line with his arm, free to move.

2. As the chain passes through his fingers, he has his hand on his hip or thigh at such a location that when the end of the chain reaches his hand, the 100-ft mark will be approximately over the stake. When the chain reaches this point, he *stops* the chain. He does not discuss the matter with the head chainman, or think about it. He just doesn't let the mark pass the stake. If the head chainman is still walking at this point, he will be stopped forcibly. A good rear chainman should never have to pull the chain back the line to his point—it should never pass the stake. The chainman is "locked on" and does not yield until he hears the call indicating that the head chainman is ready to move to the next point.

3. The rear chainman holds the chain firmly in his hand, and against his hip, thigh, shin, or ankle—depending on the appropriate elevation. Until the new stake is driven, he holds directly on his own stake, if condi-

Figure 8. The tail chainman holds the chain firmly on the mark.

tions permit; if he is on the low end of the line, he then raises the chain and plumbs.

4. When the head chainman turns, and again when necessary, the rear chainman gives him line by signaling with the right hand, putting his eye over the line. He should not relax his stance or release the chain to do this. The head chainman should not have to ask for line; when the rear chainman feels less pull in the chain, he knows line is called for and looks up. At other times he watches his point.

5. Since a chain cannot be held exactly on a point, the rear chainman calls "good" when his mark is exactly right. The lead chainman may drive a stake without hearing this call, but he knows his measurement is final only when the rear chainman announces he is exactly on the point. This is particularly important when plumbing. If the rear chainman stays on his point, he calls "good" repeatedly. This call indicates he can feel the proper pull on the tape, but he cannot tell if the sag is correct, as the pull and sag are controlled by the head chainman.

Figure 9. When not in use, the chain is slipped over the shoulders.

6. When the head chainman calls "my side" or otherwise, to indicate another point to which measurement is made, the rear chainman immediately lets go of the tape, pulls his stake, picks up his load and starts walking—or running. He gets to the next stake before the head chainman has gone the same distance, because he must again stop the head chainman.

The rear chainman should not hold the end of the chain while moving, both because it will tend to catch on obstructions if partially looped, and because he gets to the next stake well ahead of the end of the chain end. A rear chainman holding the end is a man who is delaying the work, and as one such rear chainman was told, "If you're going to ride, pick up your feet!" This refers to continuous line chaining, not layout with short measurements.

Remember, the head chainman controls the pull and sag in the chain, and the rear chainman holds firm. If the pull is too great the rear chainman can say so, but he should never give ground. The rear chainman stands with legs straight and stiff to avoid being pulled, or puts his foot flat on the ground, holding chain against bone. He never squats, as he could then be pulled off balance. If he holds the chain high, it is against his chest or head, never far from his body. The head chainman, on the other hand, is keeping a steady pull. He cannot control the position of the chain, and should keep his chain hand free of his body. The head chainman's hand is free to move if the chain moves, but the pull is kept steady. The rear chainman holds on bone, the head chainman on muscle.

In the preceding description of the head chainman's operations, a left-handed man would reverse the hand of all operations noted. The tail chainman may hold the chain in his right or left hand, as convenient, and according to how much room he has on each side. If stakes are being set permanently, they are numbered beforehand, and aligned with a transit from one end of the line; the rear chainman is careful to leave adequate room on line for the transitman to see, although he may still keep his foot on line.

BREAKING CHAIN

When the ground elevation difference between the ends of the chain is greater than 5 ft, it becomes impractical to use a plumb bob at the low end. Instead, stakes are set at shorter distances, so the distance to plumb will be less. Rather than set off several distances, such as 20 ft, 30 ft, 50 ft, and add them together, lay the chain out over the full 100 ft and set stakes at even marks on the chain. In this way, no addition is necessary and there is less chance for error.

The extra stakes set between regular 100 ft points should be furnished from those the rear chainman is carrying at the time. Since he picks them all up before moving to the next 100 ft space, count will not be lost. If the head chainman uses his own stakes, he must remember how many he has used, or there will be an error of 100 ft or more in the total distance measured.

STANDARDIZATION

Chains are checked by the U. S. Bureau of Standards for a small fee, and some engineering firms keep a standardized chain to check other chains. For most construction surveys, new tapes are considered sufficiently accurate without checking against a standard. In some areas, the city may have standardized points set in a sidewalk or a similar location for checking chains. Chains should be checked unsupported, that is, both the tape and amount of pull and sag, against a known standard, when one is available. Kinked and spliced tapes should be checked periodically to be sure they have not changed in length as a result of such damage.

CHAINING PINS

Chaining pins are steel rods about 1-ft long, with a loop in one end, painted red and white. They are sold in sets of eleven, fastened to a wire loop which is worn on the belt. When conditions are suitable, they speed up the work, as they are lighter and easier to carry than stakes, and may be shoved into the ground instead of being driven.

On construction jobs, permanent stakes are usually required, so there is little demand for pins. Some of the disadvantages of pins are listed here.

1. The point to which the measurement is made is not as definite as on a stake.
2. In soft ground they are easily accidentally moved off true location.
3. In hard or frozen ground they are difficult to place.
4. In brush they may be stepped on before they are found.

When pins are used, the chaining procedure is similar to that outlined above, but somewhat simpler, as driving the stake is eliminated.

ACCURACY

The best accuracy accomplished with the methods here outlined is about 1/10,000; that is, 0.01 ft in 100 ft, or 0.5 ft in a mile. In most cases, this accuracy will not be attained. Measurements can be repeated with less difference than this, however, because a rodman tends to repeat the same errors. If you use too much or too little pull, your error is uniform, and if the temperature does not change, temperature errors will not be discovered.

You should be able to repeat the measurement of a 1200-ft distance with a difference of 0.15 ft or less. This indicates that you are using a consistent procedure, which is adequate for training purposes when only three or four such lines are measured. Consistency, or the repetition of measurements in the same way, must be attained before accurate measurements can be made.

FIELD EXERCISE NUMBER 2

The instructor sets out a series of points (preferably marked by pipes driven into the ground) about 10 ft apart, as shown in field notes, Figure 10, to represent the one end of several lines. A second set of points for the other ends of the lines are similarly set so that each pair of pipes are about 1200 ft apart. A range rod may be placed behind each pipe to increase visibility when aligning stakes in chaining operation.

Each student chains two 1200-ft lines as head chainman and two 1200-ft lines as rear chainman. The use of a chainman's range rod is optional. If no more than two or three stakes must be plumbed on each line, this

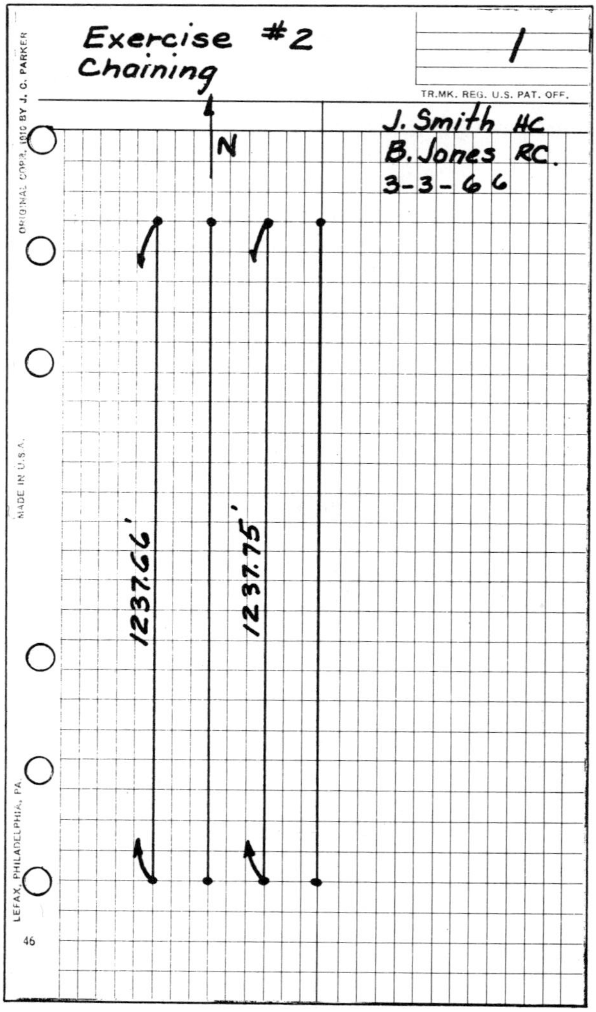

Figure 10. Notes for Exercise Number 2.

exercise should be completed by students in about two to three hours; by the time the last line is run, they will be able to run a 1200-ft line in 15 minutes.

Students form in two-man parties, each party requiring a chain, two plumb bobs, a hatchet, and eleven stakes. The head chainman is responsible for the accuracy of the lines he runs, for giving directions to the rear chainman, and for recording notes for the two lines on which he is head chainman. With range rods on both ends of the line, each party may chain one way on one line and back on another. When a suitable location is available, as along a roadside, some of the lines may be made more difficult than others.

Chapter 3

THE ENGINEER'S TRANSIT

• DESCRIPTION AND PURPOSE

The engineer's transit is the most versatile and frequently used instrument for surveying. In this chapter, *transit* refers to the engineer's instrument, Figure 11, not the builder's transit-level, which is similar but lacks many of the features of the engineering instrument.

The transit is used to measure horizontal angles between lines, to measure angles upward or downward to inclined lines, to measure approximate distances by use of sighting on a rod, to establish elevations by using the telescope as a level, and to establish directions by a compass.

Figure 12 illustrates the characteristics of a typical transit, with the lower part cut away to show the operation. In the actual instrument, the rotating parts in the leveling head cannot be seen, and the clamps and screws will differ from the drawing.

The transit consists of a telescope, fitted with internal crosshairs like a telescopic sight, and installed in side brackets so it may move up and down, even reversing itself completely, and can be clamped in any position. The brackets, called *standards,* are welded to a disk which carries the compass and is marked at the edges with *index* marks. This disk rotates to measure, or *turn,* horizontal angles.

Accessories furnished with the instrument are a plumb bob, which hangs from a hook on the bottom of the instrument to locate it over a point; a magnifying glass to read angles more easily; and adjusting pins, which are small brass pins used as wrenches, to fit into holes in the adjusting nuts.

The *telescope* has two hand turn knobs. The *objective focus knob* focuses the telescope on the target or the rod. It is usually on the top or side of the telescope, near the far end when you look into the instrument. On some instruments, this knob is on the side, projecting through a side bracket. The *eyepiece focus* ring is turned back and forth, to focus the

Figure 11. Engineer's transit. The additional scales on the vertical circle are for reading stadia distances and differences in elevation directly.

crosshairs. To adjust the telescope, this focus ring is rotated until the crosshairs appear sharp and clear; the objective is then focused on the target. It may be necessary to refocus the eyepiece later if the crosshairs have become fuzzy, and turn the objective focus knob to bring the target again into focus. The eyepiece focus should change little, if at all, when the objective focus is changed. The eyepiece focus depends on the eye of

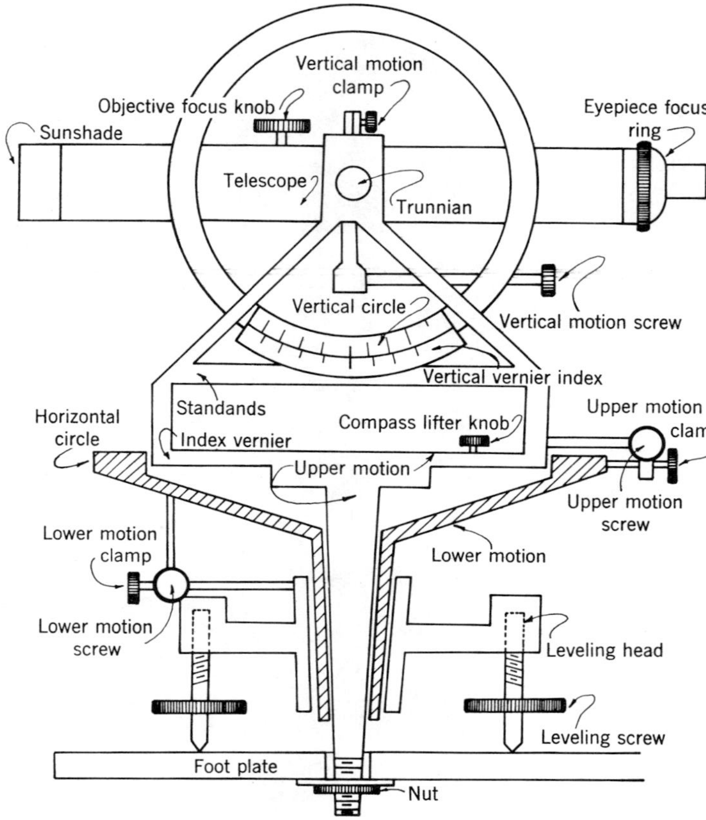

Figure 12. The engineer's transit.

the user. If you wear ordinary eyeglasses, which provide magnification but no other correction (as for astigmatism), you will not need glasses while looking through the telescope; the adjustment of the lens in the instrument compensates for the error of your eye. Some telescopes have an *eyepiece shutter,* or metal disk which covers the eyepiece when it is not in use. This shutter is opened and closed with the fingernail on a small projecting pin. The *sunshade* protects the lens of the telescope from reflecting direct sunlight back to the instrument man, which would make the image appear cloudy. When the transit is stored in its box, the sunshade is removed and a metal cover, or *dust cap,* fitted on the end of the telescope to protect the lens.

The telescope, standards, vertical circle, and compass rotate together;

this assembly is the *upper motion*. The *horizontal circle,* a ring graduated in degrees and fractions, is attached to a second rotating assembly, the *lower motion,* which is cross-hatched in the illustration, Figure 12. Movement of each of these assemblies—the telescope, upper motion, and lower motion—is controlled by a *clamp* and *motion screw*. These are pairs of knobs, mounted near each other and usually nearly identical in appearance. These upper parts of the instrument, the motions, rotate independently of each other. The horizontal circle is graduated all the way around, and it is sometimes called the graduated circle.

The clamp of each pair is like a set screw on a gear; when it is tightened, the two parts are clamped together. To move one of the two parts clamped together, the clamp screw is loosened, and then again tightened. Clamps are bolts which point directly to the center of the instrument. They do not fit into any holes, but may be tightened anywhere in their range of movement. There is a *telescope clamp* which prevents the telescope from rotating up and down in the standards; an *upper motion clamp* which tightens the upper motion, holding the telescope to the lower motion. When this upper motion clamp is tight, the reading of the horizontal circle (on the lower motion) at the index (on the upper motion) is held constant. When the upper motion clamp is tight, the telescope with both motions may still rotate freely in a horizontal direction if the lower motion clamp is loose. All clamps and screws have right hand threads. The *lower motion clamp* fastens the horizontal circle to the leveling head, and therefore to the tripod and rigid base of the instrument. If this clamp is tight, the graduated circle will not move. For example, if the circle is set to zero while the telescope is pointed north, it will remain fixed and the telescope may be rotated by releasing the upper clamp. If either upper or lower motion clamps are loose, the telescope may be rotated.

The instrument may be operated with these clamps alone. If you point the telescope on a target, however, and then tighten the clamp, the telescope will move—the least bit—while the clamp is being tightened. A *motion screw* is therefore provided for each clamp, to adjust the pointing of the instrument after the clamp is tight. That is, the clamp does not stop all possible motion of the parts, but they may be rotated a small amount by turning the motion screw either way, like a micrometer adjustment. These motion screws are mounted so that the axis of the screw is at right angles to the clamp screws. If the clamp is loose, the motion screw can be turned freely but it does not do anything; the clamp must be tight before the motion screw will move the telescope. There are three motion screws, one corresponding to each clamp: the *vertical motion screw,* the *upper motion screw,* and the *lower motion screw.*

Motion screws are also called *tangent screws.* On modern instruments, the screw is a bolt that operates against a spring; if screwed in too far, the operating distance of the spring is used up and the telescope will stop. You must then unclamp the motion, move to a slightly different location, and adjust the motion screw in the opposite direction.

The telescope rotates up and down on a pair of side shafts called *trunnions.* (The word "trunnion" was first used to refer to the same projection on a cannon.) A graduated circle, called the *vertical circle,* mounted on the side of the telescope rotates with the telescope to show the angle above or below level (the vertical angle). A modern telescope may be pointed backward by turning it upside down; this is called *flipping over* or *plunging.* To turn the transit in this way was once called to *transit,* from which the instrument got its name. Older instruments may not plunge, but are still called transits. At one time a *theodolite* was the designation for a similar instrument with a telescope too long to plunge, but this distinction is now very vague, and a theodolite may be merely a transit with more precise graduations and other improvements. Since transits and levels may remain in use for as long as sixty years, the instrument man may be called on to operate comparatively ancient variations of instruments at any time. A good older instrument is more accurate, and often easier to operate, than a cheap new one. Unlike other kinds of equipment, used transits and levels do not depreciate in value a great deal after the first few years. The chief disadvantage of an older instrument is that parts are rarely available.

The *vertical index* is a marked metal plate attached to the side bracket, which indicates the vertical angle on the vertical circle. Vertical angles are used primarily in mapping surveys, and many instrument men never need them at all. Many older transits have no vertical circle or index.

The compass needle is mounted on a pin bearing. This bearing is easily damaged, so a lever is provided to lift the needle against the glass cover of the compass when the compass is not in use. This lever is lifted by tightening the *compass lifter knob*; to avoid accidentally moving the instrument with the compass loose, the needle should be held in the lifted position when not actually in use. The compass circle graduations are not as precise as graduations on the horizontal circle, so the compass is seldom used. The angle between compass (or magnetic) north and true north varies with the section of the country and to some extent with the year. The graduated circle provided for reading of the compass angle provides the direction of the telescope; it is not graduated for pointing the zero of the circle to the north, like a hunter's or sportsman's compass. This is the reason that east of the transit compass circle is to the left of north rather than to the right, as it is on a map or other type of compass circle.

When in use, the transit is mounted on a wood tripod. The *leveling head* provides a level platform, and the *leveling screws* are adjusted to put the leveling head in the level position when the foot plate is off level. All parts of the transit are of brass or bronze, both to avoid corrosion and to avoid interference with the compass needle; its screws are soft and the threads are easily stripped. The leveling screws in particular must be carefully adjusted; they may be damaged by hand tightening. The best rule is that these screws should be as tight as can be made with the tips of forefinger and thumb. If too loose, the instrument rocks back and forth and readings are inaccurate. If too tight, the leveling head is bent, the threads stripped, and the *foot plate,* on which the screws rest, may be scarred. Proper adjustment of these screws should be checked by the instructor for each student at least once and, preferably, several times during the course.

Opposite leveling screws are adjusted at the same time. The leveling head is shifted by turning the screws so that your thumbs go toward each other or away from each other at the same time. Keeping the proper pressure on the screws is a matter of practice. Remember, the bubble on the transit plate level follows your *left* thumb.

When the leveling screws are loose, the leveling head may be shifted about $3/8$ in. on the foot plate. With this movement, you can set the transit exactly over a point without moving the tripod feet.

• BOX AND TRIPOD

The transit is usually mounted on a board which slides out from the bottom of the transit box. There may be a hand bolt fastening the board to the box, and the box door may have a keyed lock. The box has a carrying handle and, sometimes, a shoulder strap.

Tripod legs may be adjustable, so the length of legs may be changed, or they may be fixed. Adjustable legs are an advantage on rough ground, or in making a setup on the side of a ditch; the uphill leg is shortened and the two downhill legs are extended to full length. The adjustable leg tripod is also easily packed in a car. Do not tighten the screws on adjustable legs when the tripod is put away. If you do, the legs may expand due to the humidity in the air, which tightens the screws so that they can not be loosened again without breaking.

• SETTING UP THE TRANSIT

First, remove the protective cap from the threads and set up the tripod, extending the legs if necessary.

Open the box and slide the transit out on its board. Hold the instrument cradled in one arm and hand, against the body, as a mother holds a young baby, with the board in the direction of the baby's head. Hold the leveling screws in the hand supporting the instrument, and unscrew the board with the free hand. The transit should never be picked up by the telescope when this can be avoided, but some transits are stored in top-door boxes and must be handled in this way. If handling by the standards or by the telescope, be careful that you do not twist the side brackets by lifting on one side more than the other.

Figure 13*a*. How to carry the instrument the right way.

Loosen the upper motion clamp to allow rotation of the base plate. Supporting the instrument by the standards or by the upper plate, set it on the tripod. Continuing to hold it with one hand, rotate the foot plate backward until a click is felt. This indicates the thread end of the transit has just passed the thread end on the tripod. Then rotate the foot plate clockwise, tightening the threads firmly on the tripod. If the ends of the thread are not located in this way and the instrument is not held perfectly plumb, the threads may cross and damage each other.

When carrying the instrument mounted on the tripod, keep the tripod feet together, and held by one hand. The arm balances the instrument on the shoulder. The points of the tripod legs should point forward, so if a person stumbles he can push the feet of the tripod into the ground as he falls, reducing the chance of damage to the instrument. The telescope should be loosely clamped so that it will give if it hits an obstruction. When passing through brush, under trees, or through a door, the instrument and tripod are carried crosswise in front of the chest, and held with both hands. In this position, they are unlikely to hook onto obstructions. The shoulder carrying position is normal, and should be used by the

Figure 13*b*. How to carry the instrument the wrong way. Don't fight the job by spreading the legs.

student even if moving only a few feet. On the shoulder, the instrument requires but one hand, and can be carried for miles; an experienced man assumes this position automatically. The new man will be considered an amateur if he attempts to carry an instrument even a very short distance with the arms spread or partially spread, or with the instrument held at arms' length in front of his body as though he had caught an octopus by the head and was trying to avoid the tentacles!

Install the sunshade before work is started. The magnifying glass is best

Figure 13*c*. How to carry the instrument the wrong way. Even for a short distance, learn to carry the instrument the easy way.

carried on a string around the neck, especially for student use, and when working on jobs where it will be used frequently. If carried in the pocket, the glass is less available and frequently will be forgotten at the end of the day. While working with the magnifying glass without a string, it is often left on top of the transit plate when not in use; the transit is picked up, and the glass dumped in the mud or in the grass. The plumb bob may be carried in a holster or in the pocket, but in the pocket the point not only penetrates the trouser leg, but also punches a hole in the hand if it brushes the thigh, with painful results. When moving the transit a short distance, the plumb bob need not be removed from the instrument, but merely dropped in the holster.

Under ordinary circumstances, the transit box should be left in the tool shed or truck. It is easier to carry the instrument on the tripod, and the instrument is more resistant to weather than is the box. When the transit-man reaches his point, he spreads the legs of the tripod as nearly equally each way from the point as he can estimate, and sets the points of the legs slightly into the ground. The plumb bob string is looped to the hook on the bottom of the transit, and fastened with a slip knot. The string should be of the same type as the plumb bob string used for chaining. A heavier plumb bob usually is used for the transit than that used for chaining, because it is not carried or used so much and the heavier bob is more

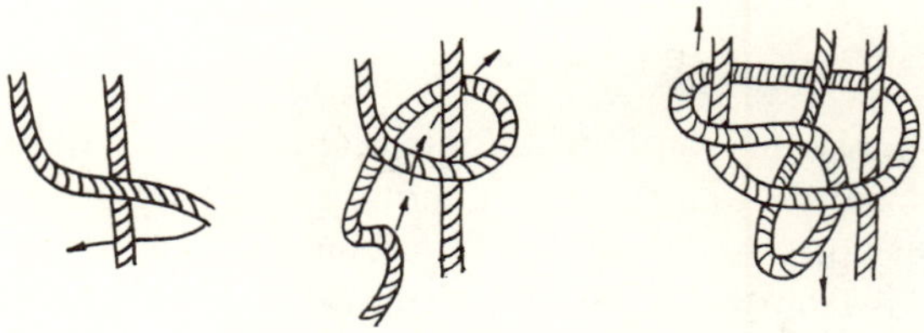

Figure 14. The slip knot.

steady. If the transit is originally provided with a light plumb bob, you will very likely want to replace it with a heavier one. The plumb bob holster is also necessary for the instrument man, but this is not provided with the instrument.

Using the slip knot, adjust the plumb bob up or down without changing the tightness of the knot; to tighten the knot, pull on the top. If the S-hook under the transit is lost, a paper clip can be bent, and used as a substitute. The plumb bob string is replaced by removing the threaded cap, inserting the string, knotting it inside the cap, and replacing the cap. There should be no knots in the string except at the end, or the slip knot cannot move freely. Part of the equipment for a surveying crew should include extra string for periodic replacement of knotted or unraveling string. To remove the plumb bob from the transit, pull out the slip knot with a single yank. There are patented devices available which fit onto the plumb bob string, replacing the slip knot. It is this writer's experience that any time gained through the use of such devices is lost by tangling of the string when carrying the instrument or the plumb bob.

The instrument is roughly located over the point by adjusting the location of the tripod feet, remembering that the instrument moves in the direction of the foot being shifted, but only half as far. After the feet are firmly forced into the ground, the final adjustment is made by shifting the instrument on its foot plate. The plumb bob should almost touch the point when final adjustment is made; do not leave it high and then guess where the center below falls. Some instruments have a special telescope device through which the surveyor can look at the point below, rather than using a plumb bob. This *optical plummet* reduces the error in plumbing, particularly when the wind strongly affects the string.

The instrument is roughly leveled when the leveling screws are loosened to move the instrument on the foot plate; after the center of the instrument is in the exact location, the instrument is leveled in both directions by adjusting each pair of screws by the "bubble follows left thumb" rule mentioned before. You are now ready to go to work.

THE HORIZONTAL CIRCLE

The angle is indicated both by number and marked divisions on the outer ring (horizontal circle on lower motion), and by the zero mark on the inner index (upper motion). The horizontal circle is usually marked in both directions from zero; clockwise from 0 to 360°, and counterclockwise from 0 to 180°, starting over again and running to 180°, but instruments vary. Since the angle can always be read clockwise, all references in this textbook are to clockwise readings, or to angles turned in a clockwise direction. Transits are graduated so that the finest reading is 1′, 30″, or 20″. If an even degree, or a fractional degree which is graduated on the circle, is being read or laid off, the index zero line will appear to be a continuation of the mark on the graduated circle. If the index mark falls between the graduations on the outer circle, the vernier line that corresponds, as shown in Figure 15, will show the reading. The numbers on the vernier will correspond to the finest graduation on the circle; that is, if the circle is graduated to thirds of degrees (20′), the numbered marks on the vernier will go up to 20. The illustrated vernier is graduated to read only to one minute. If finer graduations appear, they will be subdivisions of the minute marks on the vernier; for example, the 20 marks shown may each be subdivided into three parts, so that an angle may be read to the nearest 20″.

The instructor should build a vernier board for class use; because of the small size of a transit vernier, it is difficult to explain transit graduations to more than one person at a time.

ERRORS IN READING ANGLES

As you are reading angles clockwise, you will always use the left-hand half of the vernier. The right-hand half indicates the minutes over the

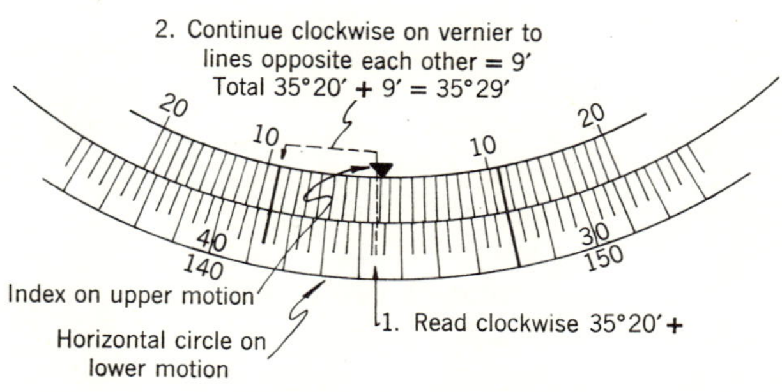

Figure 15. Reading the horizontal angle.

third-degree mark in the other direction; in Figure 15 this difference is 11′. Or from another standpoint, since the index is 9′ past one 20′ mark, it is 11′ short of the next one. Since there will be a mark on the right-hand side of the vernier that will correspond to a circle mark, as well as one on the left-hand side, it is important to read the left-hand side only.

An error often made by beginners is to use the line in the *graduated circle* that corresponds to the vernier line as a reading. There is no relation between them. For example, the 38°40′ mark on the horizontal circle in Figure 15 has no meaning; the 9′ mark above is read in connection with the 35°20′ mark on the graduated circle.

VERNIERS A AND B

To minimize errors, engineering transits are equipped with vernier markings on opposite sides of the upper motion; one is designated *A*, and the other *B*. In precise work, both of these are read, one set of readings adjusted for the 180° difference between them, and the average values used. This eliminates errors of calibration of the graduated circle, but is not necessary for ordinary work.

LAYING OFF A GIVEN ANGLE

If you set up a point and are directed to turn the telescope a certain angle from a point already set, in order to set another point, you are laying off an angle. This occurs in construction work, as when laying out a

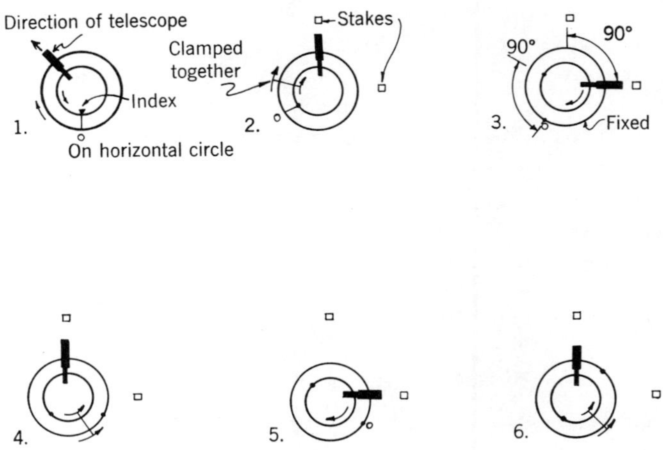

Figure 16. Angle measurement by repetition.

building or a centerline of a highway, or in land surveying when running the boundaries of a new lot.

The index on the horizontal circle is first set at zero, as in step 1 of Figure 16, clamping the upper motion clamp. The telescope is pointed toward the known mark and set on it with the lower motion clamp, as in step 2. The lower motion clamp is left tight; but releasing the upper motion clamp, the telescope is pointed so that the given angle is read on the horizontal circle. The upper motion clamp and screw are used to set the reading exactly. The telescope is now pointed in the desired direction to lay off the new line.

REPEATING ANGLES

The angle to points in place can be determined more accurately than a line can be set, since the angle may be "added up" on the graduated circle several times, and the total reading divided to measure the angle. For example, on an instrument reading to 20″, the most accurate reading which may be obtained will be to the nearest 20″. But if you read the total of three angles which have been added up on the graduated circle, you can still read the total to 20″. You divide the total by three, which will give the average to the nearest ±6.7″. This is true if there is no error in pointing the telescope or in setting up over the point, but there are such errors in short distances. For example, if on a 100-ft building the transit is set up 0.005′ (1/16″) off the correct location, and there is an error of the same amount in setting on the target (1/32″ on each nail), the total error will be 0.01′, which at that distance amounts to 20″ of angle. Repeating angles on such a short distance is convenient for student exercises, but on an actual job there is little accuracy gained by repeating angles if distances are under 400 ft.

The procedure used in repeating angles is illustrated in Figure 16. The angle may be repeated as often as desired; some kinds of work require an angle to be read at 10, 20, and 30 repetitions. These are on long sights, often over ten miles, and are made at night when there is less distortion in the atmosphere. For ordinary work, three readings are sufficient. Each angle is read as a check on the work, but only the last reading is used for averaging. For example,

First reading, 90°00′.
Second reading, 180°02′ (difference 90–02).
Third reading, 270–03–20 (difference 90–01).

The first two readings need be made only to the nearest minute, as they are for the purpose of detecting errors in pointing the instrument. If the difference between successive readings is more than 1′ over or under the average difference, the transitman learns that he has made an error in pointing or in the original setting, and starts over at the beginning. If a motion screw is accidentally turned a small amount, or if there is an angle other than zero on the graduated circle at the beginning, the difference between successive readings will reveal this error.

CLOSING THE HORIZON

When all angles at one setup are measured and form a complete circle, the instrument man *closes the horizon.* Although seldom done in ordinary work, it serves as a check on the angles, since they should add up to 360°. Field notes for measuring angles are shown in Figure 17.

"All angles direct" refers to the fact that the telescope was in the up-

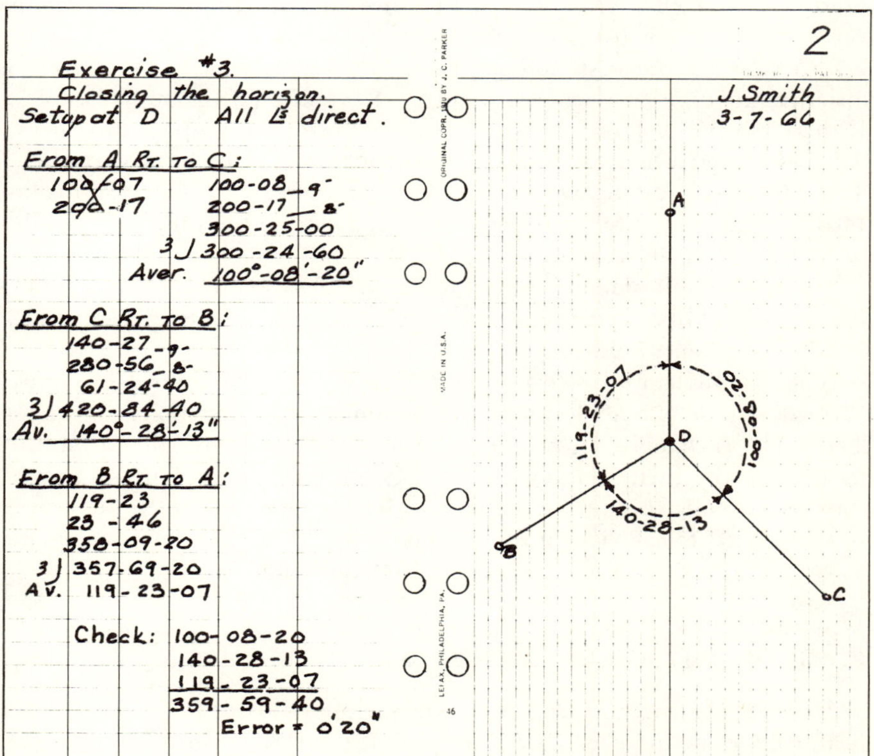

Figure 17. Field notes—closing the horizon.

right, or normal, position. For precise work, both angles are turned this way, with the telescope flipped over, in the *reverse* position. This eliminates errors caused by the telescope moving up and down in a direction not perfectly plumb, with targets at different elevations.

The first two angles measured in these notes were crossed out, because the difference between the first and second readings, 100–10, was too far off the first reading, 100–07. This difference indicated either a faulty reading or an error in setting the instrument. The second set of angles showed that the angle read as 100–07 should have been 100–08, which brought the differences between permissible limits.

The small number between each pair of readings is the last number of the differences between the successive readings. For example, in the second set of angles, from *C* right to *B*, the difference between the first and second readings is 140–29, but it is necessary to check only the last digit, or 9. Note that when the accumulated total passes 360°, the actual reading of 61° – 24′ – 40″ is recorded from the instrument, although the angle is 360° greater. When using this method, which saves a little work, remember to subtract the first number from the last, not vice versa. In this instance, the difference between 7 (of the 27) and 6 (of the 56) is 9, not 1. If there is a 10′ or larger, error, which would not show up in this check, it is discovered when the final angle is compared with the first reading. The 140–28–13 for a final angle is compared with the first reading of 140–27. Had there been a 10′ error in any reading, the difference would be at least 3′ rather than 1′.

The degree, minute, and second marks are often omitted in field notes, but zero should be used when there are less than ten units. For example, 7°7′7″ is written 07–07–07.

When an angle must be divided by three, as in the field notes in Figure 17, first rewrite the angle in divisible numbers. For example, the angle 300°25′00″ is changed, in two steps, to 300–24–60 by converting a minute to seconds, and then dividing by three. If you try to "carry" minutes over into the seconds column mentally, an error is very probable. The mind is accustomed to carrying in the decimal system, and it is very difficult to learn to carry alternating decimal and 60-unit numbers.

CARE OF THE INSTRUMENT

A transit is a rugged instrument when properly used. It will withstand heat, cold, rain, and even immersion in salt water. But since it is made of soft noncorrosive metal, nearly any part can be damaged with the bare hands; some parts may be damaged by the pressure of two fingers. The pressure applied to the motion clamps and the leveling screws must be light but firm; the tips of the thumb and forefinger are sufficient. Oil should be applied sparingly or not at all; working parts should be lubricated with graphite when needed. Oil collects dust, and in cold weather becomes sticky.

The leveling screws should be less tight when the instrument is stored than when it is in use, but like all clamps they should not be so loose that parts of the instrument can rattle. When the instrument is stored in the box, motion clamps should be just loose enough so that the telescope will center itself in the brackets provided. In some boxes, the leveling screws must be loosened and the instrument placed back on the foot plate as far as it will go, or the box door cannot be closed. Never force the instrument into the box, or force the door closed; if it does not fit properly, the instrument is not properly inserted in the box.

It is very important that the wing nuts on the tripod legs be loose when the instrument is stored, because of possible expansion of wood legs. Straight-leg tripods have such wing nuts at the tripod head, and adjustable-leg tripods have them on the leg extensions as well. A weak point of the tripod is the attachment of the metal tips to the wood legs. There are small projections attached to the tips, which, when stepped on, force the tips into the ground. Do not force the projections by kicking, or they may break off.

If the instrument is set up on a hardwood or smooth concrete surface, the legs may slip. The transitman should have a triangular wood frame to set the legs in, if he is to work in such locations. A piece of sash cord will also serve as a safety frame to prevent spreading of the legs. Be careful that the setup is not made so high, particularly on highways, that the instrument can be blown over by the wind. There is a temptation to make such a high setup because of the danger to a normal setup from passing cars. Remember to put the legs of the tripod in such directions that a leg does not interfere with the line along which measurements are to be made.

It is common practice to paint tripod legs alternately red and white, like the range rod, to increase visibility. This is a safety feature on the highway, and is helpful in other locations for the rodman to line up with the transit.

A transit is a valuable, light, and easily pawned tool, so you should not leave it unattended anywhere you would not leave your watch, portable radio, typewriter, or wallet. Transits have been stolen from a highway while the crews have been looking directly at them.

CARE OF THE INSTRUMENT MAN

When you are transitman, stand in a comfortable position; it is inefficient to bend your back and lean forward when reading the instrument. Your feet should be spread widely apart so that your eye is at the instrument level as shown in Figure 18. A man over 5 ft, 8 in. tall may have to

Figure 18*a*. How to stand at the instrument. This man is in a comfortable position, both eyes open.

lean forward, but he should keep his back straight and support himself by an arm on his knee. Attempting to stand up beside the instrument, twisting the head sidewise, is a mark of an amateur. Since a man will not keep this position very long, the rodman will have to yell or wave a flag every time line is needed. The purpose of a proper stance is to keep the transitman on his job; this cannot be done if he adopts a cramped position. Remember, don't stand in a position you cannot remain in all day!

For the same reason, you should learn to keep both eyes open when looking through the telescope and magnifying glass. You can then locate targets in the field of view more easily, as one eye sees the crosshairs and the other eye the target, thus preventing eyestrain. If one eye is kept closed, it becomes painful in a very short time. Looking through a transit is a full-time job, and you should not expect to look occasionally, divert-

Figure 18*b*. How not to stand at the instrument. This man is hunched and uncomfortable.

ing your attention to other matters the rest of the time, any more than you should expect a taxi driver to look at the road once in a while.

Sunglasses, like other glasses, and hard-brimmed hats, interfere with reading an instrument. A soft-brim hat is recommended, except where safety considerations require a safety helmet. Some instrument men use a cap, worn reversed when at work.

It is essential that nothing touch the instrument tripod during the work, and the transitman's feet should be as far from the tips of the tripod legs as possible. When setting up on soft ground, it may be impossible to walk near a transit without disturbing the readings, and special provision must be made for supporting it.

FIELD EXERCISE NUMBER 3

Each student needs two or three stakes, with nails and a hatchet, and a transit. Set up stakes as shown in Figure 17, about 100 ft away from the instrument setup in three directions, or as far as space permits. Since a man at one setup may use some of his neighbor's stakes, each man does not need a complete set. If the site is shady or the day dark, it may be necessary to attach bits of paper to nails in stakes to see them; usually, nails in the top of the stake, projecting at least half an inch, will serve for sights.

Measure each of the three angles by triple repetition, as shown in the notes. When you clearly understand how to read the vernier, work will go rapidly; the last set of triple angles should be readily completed by a student in 25 minutes. The exercise as a whole will require about three hours for slower students, and for those who have to repeat parts of the work. The total of all angles should add up to within a minute of 360° to be acceptable; for many students, the total will be within 10″.

The most common error a student makes is to repeat and read the angles but fail to check each difference to see if he is reading the angles right. In this way, he continues to read angles incorrectly, and must go back to the beginning to correct his errors.

The instructor should check all instruments as soon as they are set up to see that the leveling head screws have the proper tension, and then check the faulty setups a second time. He should check each student as soon as the second reading has been made on the first angle, to see that the student is reading the vernier properly, and has computed the difference between the first and second readings.

It is important that the student learn at this stage that he must record all data immediately, and must make all calculations promptly, and on the spot, to avoid having to repeat work. This is particularly important in later exercises. The student who attempts to put off the portion of the exercises he understands least is not only wasting his field time, but may be acquiring faulty procedures as well. He is learning less than nothing.

Chapter 4

THE LEVEL

A large part of the engineering surveyor's work is to determine elevations. Land elevations, water elevations, and elevations of previous construction comprise the greater part of preliminary surveys. The construction engineer repeatedly sets marks indicating elevation. Initially, on a building, these are heights of batterboards. After the first floor is completed from the batterboards, he must measure up to set elevation marks at other points on the site or in the building.

Each floor slab, in a multistory building, must be laid to new marks, and checked with a level for each floor. Pipelines are staked to indicate depth of cut and elevation of pipe, usually at 50-ft intervals along the line, and site grading must be repeatedly checked to assure proper drainage.

LEVEL SURFACE

For our purposes, we may assume that a level surface is a flat plane, that the earth is flat, and that the surface of a body of water at rest is also a flat plane. The curvature of the earth requires that a level surface be curved. The procedures used eliminate this effect, and this curvature is measurable with ordinary instruments.

It is also assumed that all plumb bob lines (that is, vertical lines) are parallel to each other, rather than coming to a point in the center of the earth. This effect is substantially true. In one of the largest structures ever built, the bridge over the entrance of New York harbor to Staten Island, the distance between the tops of supporting towers is but $1\frac{1}{2}$ in. greater than the distance between their bases, caused by the curvature of the earth.

In surveying of extreme precision, such as for intercontinental missile firings, and for locating astronomical observatories on opposite sides of the earth, the fact that the force of gravity is not toward the center of the earth must be taken into account. A plumb bob actually points toward the center of mass of the earth, as viewed from that particular location, and shows a tendency to be diverted from the lighter parts of the earth toward the heavier portions.

REFRACTION

It is convenient for most purposes to assume that light travels in a straight line; only rarely does it substantially bend in air on the earth's surface. Bending of light when it enters a glass of water is a well-known phenomenon. Light from the stars and sun also bends when it enters the earth's atmosphere; this bending is called *refraction*. Because of this effect, the sun appears to come up suddenly, and to set quickly, and the day is longer than is indicated by the assumption that light travels in a straight line.

In surveying, a line of sight may be bent by passing from warm to cool air, and vice versa; this bending may be as much as 1 ft in a mile. When precise work is necessary, it is best done on a cloudy day or at night, when there is little air movement or difference in air temperature.

METHODS OF DETERMINING LEVEL LINES

Levels are determined by some method of laying off a line at right angles to a vertical axis (that is, at right angles to the direction of the pull of gravity). If a level line is used to determine the elevation of two points that are not the same distance from the point where the level line is determined, the curvature of the earth causes an error of about 0.6 ft in a mile, or 0.001 ft, which is measurable on a level rod, in 200 ft. In working from one setup, this can be ignored, but it is taken into account when measuring the difference in elevation between points a considerable distance apart.

There are three commonly used methods of determining a level line.

1. You can hang a plumb bob or other kind of weight to determine the direction of gravity, and use a fixed mechanical device to measure the 90° angle from the vertical to the level line. The most ancient such device, of which one variety is shown in Figure 19, is the chorobate. This consists of a triangular frame, with a mark on the lower horizontal bar to show the proper location of the plumb bob when the lower bar is level.

Variations of this device were used for thousands of years, and were quite satisfactory for short distances. It was the only leveling device available for projects like the cutting of a canal from the Nile to the Red Sea, and the Roman attempt to cut a canal across the Isthmus of Corinth in Greece, which was stopped, according to one story, when the engineer

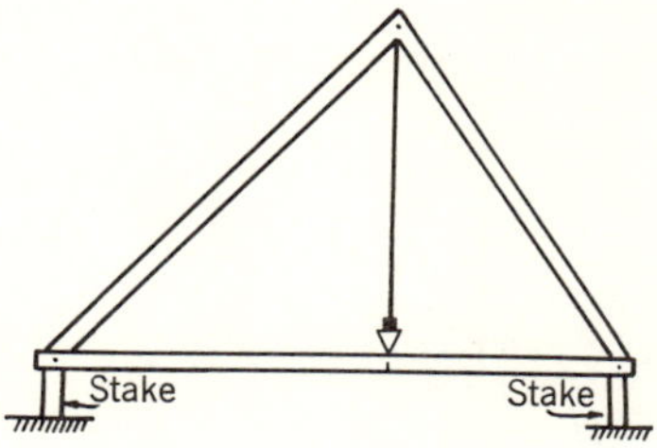

Figure 19. Chorobate.

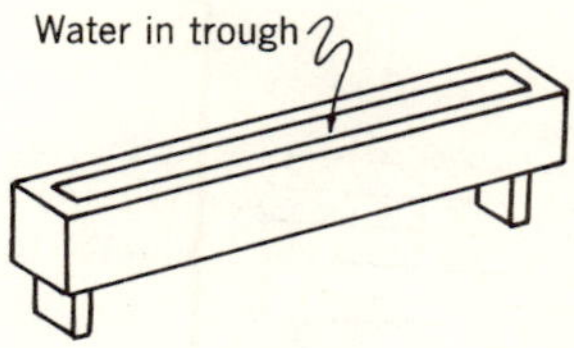

Figure 20. Grooved plank.

erroneously reported that the difference in water elevation made the completion impractical.

This device, like all others, may be checked by reversal. That is, if two stakes are set at the same elevation with the chorobate, and the instrument is reversed and still shows the stakes to be at the same elevation, it is in adjustment. If the plumb bob does not return to the same point when the instrument is reversed, the instrument is in error.

This principle was abandoned in the manufacture of leveling instruments until the last twenty years, with the acceptance of the "self-leveling," or automatic leveling, telescope. This instrument depends on the force of gravity to keep the line of sight horizontal, rather than on a manual adjustment of a spirit level.

2. The tendency of water to seek its own level is used in another method. The device illustrated in Figure 20, a grooved plank with water in the groove, was well known to ancient builders. Archimedes complained that it was not strictly accurate, because of the curvature of the earth, which is true, of course, of any leveling instrument.

In modern times, this principle is used in the lather's water level. This is a hose full of water (as shown in Figure 21) with a glass tube in each end. The water level in one tube will always be the same elevation as in the other. One tube is therefore held against a mark as a starting point, and another mark is made at the water level in the other tube at another location. This is an accurate tool, particularly for transferring elevations around corners, and may be used for building batterboard elevations as well as for ceilings.

3. The last method of determining a level line uses the fact that an air bubble will seek the highest point of a curved tube of liquid. Both a curved tube and a partially spherical glass lens are used. The accuracy of this type of level depends on the precision of the radius to which the interior surface of the glass tube is manufactured. In engineering levels, this radius may be several hundred feet.

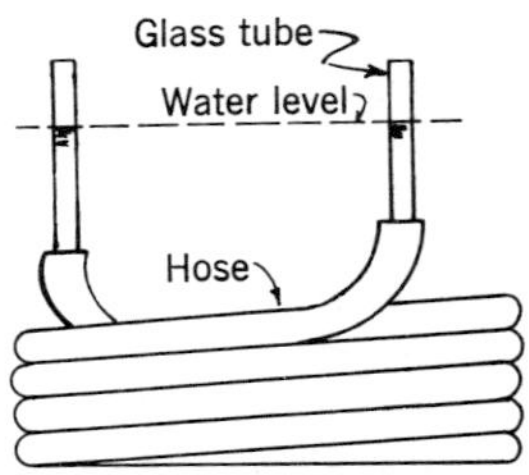

Figure 21. Lather's water level.

Levels using a bubble are called spirit levels, and are used in all telescopic levels, except the self-leveling type. In the self-leveling level, there is a spirit level but it is used for the initial setup of the instrument. "Spirit" refers to the alcohol used as the liquid.

SPIRIT LEVEL INSTRUMENTS

The *carpenter's level* (Figure 22) is a part of every carpenter's tool kit. It may be of wood or of aluminum, and is provided with at least four spirit levels—sometimes more. Each side of the tool can be directly checked to see if it is either level or plumb. The long sides are machined straight and parallel, and the ends are square. Bricklayers use a similar level, but it is somewhat longer. The carpenter prefers a tool short enough to carry in his tool box, but a bricklayer has his other tools in a canvas bag, carrying the level separately.

To make a longer tool, a level may be made of any straight piece of wood. Small spirit levels may be attached directly to a board, and such a level, often called a straightedge, is as accurate as the lumber is straight (Figure 23).

The carpenter's level may be used with a plank to set batterboards, or may be clamped to a post and used as a sighting instrument for approximate work.

The *string level* (Figure 24) is an inexpensive spirit level about 4-in. long, made of a glass tube in a sheetmetal frame. The bottom is flat, so it may be used like a carpenter's level, and each end is provided with a hook to attach to a string. This tool is very inexpensive. The bubble is not accurate in comparison with other types, and the string is not straight

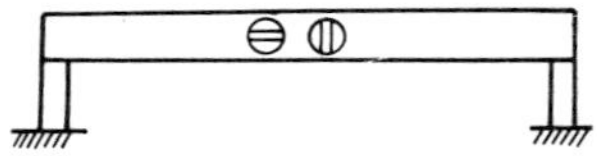

Figure 22. Carpenter's spirit level.

Figure 23. Straightedge with attached level.

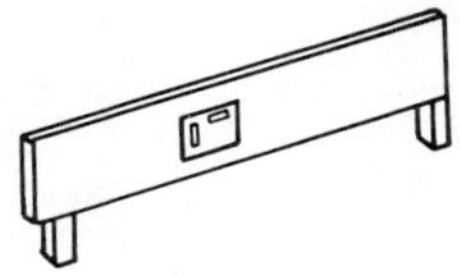

(regardless how tightly it is stretched), nor is it of uniform diameter. This tool is therefore not sufficiently accurate for most work, but it is useful for transferring elevations from grade stakes for excavation work. By careful repetition, a house 30 ft square can be batterboarded with this level, with a maximum variation either way of $\frac{3}{8}$ in. With fine wire rather than string, sag is reduced, but the bubble accuracy limitation is still important. The string level must be placed in the middle of the string, which requires one man to manipulate the end of the string and one to read the level.

The *hand level,* one variation of which is called the Locke level, is a sighting tube so arranged that when held in the hand the operator can see both the bubble and the level rod. The bubble appears in one half of the tube, reflected in a mirror, while the other half is open, enabling the operator to see the rod (Figure 25). It does not magnify, as does a telescope.

To sight on a rod with the hand level, the centerline of the spirit level, which appears in the mirror as a horizontal mark, is read on the rod at the same instant that the bubble is centered on the same mark. Accuracy is limited by the distance from which rod markings may be read directly, so the standard level rod with graduations of 0.01 ft may be read to these units only to a distance of about 30 ft; and since the level has no support, accuracy is limited by the operator's ability to hold it steady.

The chief restriction on accuracy is the surveyor's ability to hold the level still long enough to read it, as the bubble and rod must be read with a different focus of the eye. Hand-supported, an accuracy of 0.1 ft in 30 ft is reasonable. If supported on a post and the measurement repeated several times, the error at this distance can be reduced to about 0.03 ft.

The hand level is used to determine approximate cut and fill for rough measurements of grade, and sometimes in chaining to determine the difference between the elevation of the ends of the tape. It may also be used for setting rough grades for excavation.

Figure 24. String level.

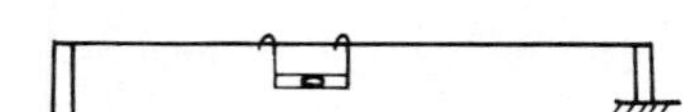

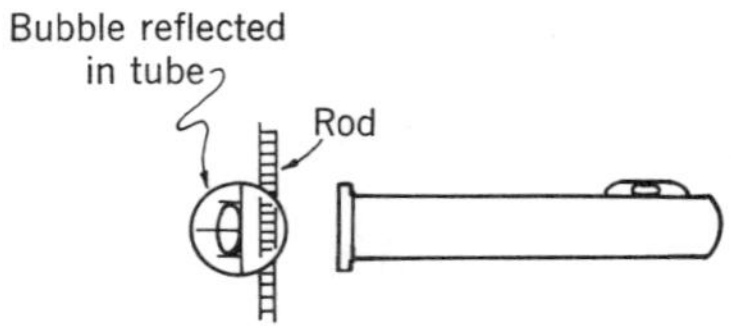

Figure 25. Hand level (Locke level).

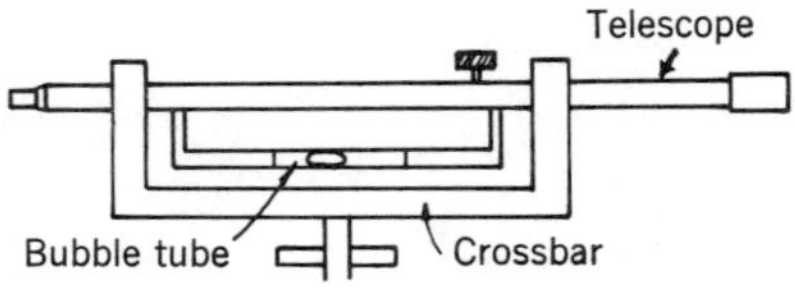

Figure 26. Telescope level.

TELESCOPIC LEVELS

For accurate work at longer distances, telescopic levels are used (Figure 26). The essential parts are a telescope with a horizontal crosshair, a spirit level attached to the telescope to show when the telescope is level, and a graduated rod, held vertically to measure difference in elevation.

Telescopes are attached to the frame in one of three ways.

1. As in *wye levels* (Figure 27), the telescope rests in open brackets, or *wyes,* which may be opened to remove the telescope. By adjustment of capstan nuts, the brackets may be raised or lowered on the frame.
2. As in *dumpy levels* (Figure 28), the telescope is rigidly welded to the frame with no provision for adjustment or removal.
3. As in *transits* and *transit-levels,* the telescope is supported so that it

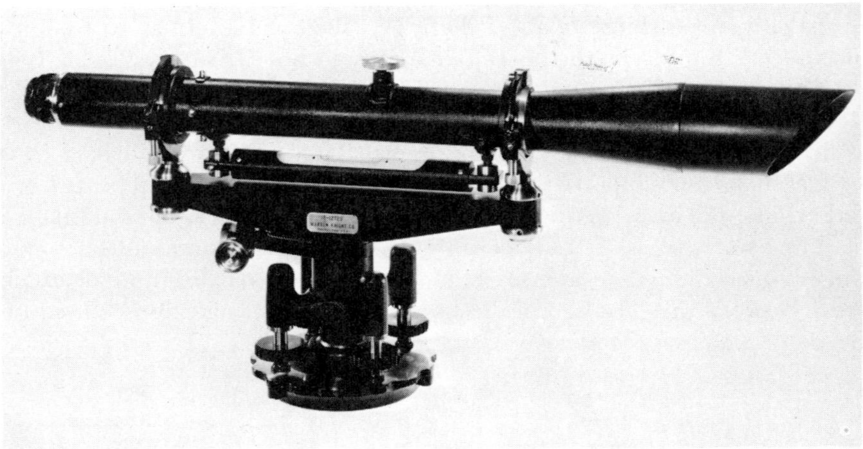

Figure 27. Wye level. (Warren-Knight)

Figure 28. Engineer's dumpy level. (Dietzgen)

may rotate up and down, being locked or clamped when used as a level A transit-level is an inexpensive level that has a restricted vertical motion.

WYE LEVELS

It is said that the wye level is the easiest instrument to adjust, and usually needs it. It is not so popular as in past years, but is still widely used. In 1965, about a tenth as many wye levels as dumpy levels were manufactured.

The telescope is removed from the frame by releasing clips at the top of the telescope, and is often packed in the box separately. The telescope fits into wyes so that it may be rotated without changing the line of sight, and it may be reversed in the supports, end for end, without changing the level of the telescope. By turning the frame nuts on the wyes, the bubble tube may be adjusted to a true right angle to the vertical axis; for this reason, it is always a simple operation to adjust the wye level so that it will rotate while in use and still remain level. Dumpy levels lack this adjustment.

Special methods of wye-level adjustment are available in standard textbooks on surveying. Because of lack of experience with wye levels, many younger engineers choose the dumpy level. Most of the old wye levels in use have a more sensitive bubble than the usual dumpy level. The bubble tube on the wye level may be adjusted in the same way as described in this manual for other bubble-type instruments.

DUMPY LEVEL

Nearly a hundred years ago, wye levels were used by American engineers, and fixed levels by the English. In accordance with European practice, English levels were *inverting type,* which meant that the image appeared upside down. Since such construction saved the use of two lenses, these levels were shorter than our standard type. Because English levels were shorter, they were called "dumpy" in this country, and when the same type of fixed-frame level was manufactured in this country in later years, it was actually made the same length as the wye level for the same magnification, but the name "dumpy" remained. The term now refers to a telescope attached to the frame, without regard to the length of the telescope.

The dumpy level stays in adjustment better than does the wye, but in order to make it rotate in a level plane, a crosshair adjustment must be made to correct for any distortion in the frame. Although distortion does not frequently occur, it is a more difficult adjustment. When buying a dumpy level, it is important to get one with a rigid joint between vertical axis and frame; some instruments have a screwed joint, rather than a soldered one. An instrument repairman can advise you on this.

Builder's levels (Figure 29), intended for use at short ranges, are of somewhat lower cost and accuracy. Some are of quite high quality, comparable to engineer's dumpy levels; prices vary from $450 to $50. The sensitivity of the bubble is less than it is in engineering instruments; the

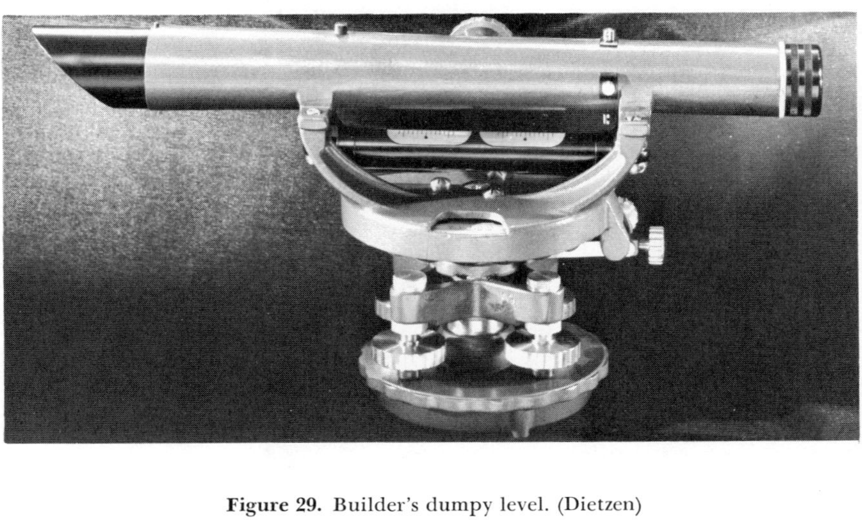

Figure 29. Builder's dumpy level. (Dietzen)

best builder's levels will give results to within 0.01 ft at 100 ft, and the cheapest of them within 0.04 ft at 100 ft. These instruments, rather than the transit-levels, are usually built to operate as dumpy levels.

TRANSITS AND TRANSIT-LEVELS

The engineer's transit has a fair accuracy when used as a level. Since the telescope and bubble tube must be shorter, it cannot have the accuracy of a longer instrument, but is equal or better to a builder's level of similar length.

Transit-levels (Figure 30) are builder's levels that are built so that the telescope may be released for use as a transit. They are less accurate than the engineer's transit for measuring horizontal angles, and have a limited vertical movement of the telescope. The telescope is too long to plunge or reverse itself. One type of builder's level has a removable telescope that is set in wyes as a level, and in trunnions as a transit. It is important that builder's levels be used so as to minimize the error at any one place; in resetting grade stakes, the same locations for setup should be used so that errors from one direction do not accumulate and show up as a single

Figure 30. Builder's transit-level. (Dietzen)

large error where two lines from different setups come together. The insensitivity of the bubble makes shots over 100 ft undesirable. Many contractors prefer a used engineer's level to a new builder's instrument at the same price. However, the difference between the best and the most expensive builder's levels, and an engineer's level, is not significant; and there is little difference in prices.

USING THE TRANSIT AS A LEVEL

When used for setting grade stakes, or for other work requiring many *shots* (readings) from each location, the transit requires more care in setting up if it is to rotate with the telescope remaining level. This adjustment is for speed and convenience; releveling the instrument for each reading does not affect the accuracy of the work. The popular idea, implied in some engineering texts, that a level will not do accurate work merely because the bubble goes off center when the instrument is rotating, is *not* true.

To set up a transit so that the telescope will rotate and remain level, there are four steps.

1. Set up the instrument with the plate bubbles and telescope bubble centered. This setup would be sufficient if the plate bubbles were sensitive enough to make the vertical axis truly plumb. The telescope bubble, but not the plate bubbles, have this sensitivity.
2. Rotate the telescope 180° on the vertical axis. If the telescope bubble is centered, no further adjustment in this direction is needed. If the bubble of the telescope goes off, bring the bubble *halfway* back to center by use of the slow motion screw on the telescope vertical movement, and bring the bubble back to center the rest of the way by use of the base plate leveling screws. This adjustment should be so small that the bubble on the base plate is not affected.
3. Rotate the telescope another 180° to bring it back to the starting point. The telescope bubble should be centered; if not, repeat step 2 and 3, if necessary.
4. When the bubble is stable, or nearly so, in the original direction, repeat the above steps at 90° to the original adjustments, and recheck the first direction used.

This process is making a setup, not adjusting the instrument. The transit bubbles must be in correct adjustment before attempting to use the transit as a level.

This setup is comparatively time-consuming, although the adjustment will remain, held by the telescope motion clamp, until disturbed. This telescope position will usually be changed when packing the transit in the box. You will not usually make such a careful setup unless a large number of readings are to be made from one location.

DIFFERENTIAL LEVELING

Elevations are determined by comparison of one point with another one nearby. The point of reference, called a *bench mark,* is a permanent point of known or assumed elevation. On a construction job, the owner's engineer often sets these points, and the contractor works from them. If there is no reference mark set when the preliminary survey is made, as is often the case on a building survey, the architect will indicate some point of reference, even though it may be a specification requirement that "floor shall be one foot above street curb." In this case the curb, at a specified point, becomes the engineer's bench mark.

The bench mark may be a square marked on a curb, a nail driven into a tree, or a similar point. To determine elevations, a graduated rod is held on the point, and by means of the telescope a measurement is made upward to the level of the instrument.

This measurement is made by reading the rod directly with the telescope, at the intersection of the crosshair with the figures on the rod (Figure 31). Such a rod, the common type, is called a self-reading rod. The vertical hair in the level should be just off the rod for this reading; if on the rod, it may interfere with the reading and if too far off it, some error may be introduced if the horizontal hair is not exactly horizontal, or if there is distortion in the instrument.

This first measurement is a *plus* reading, as it is *added* to the elevation of the bench mark to determine the elevation of the center of the instru-

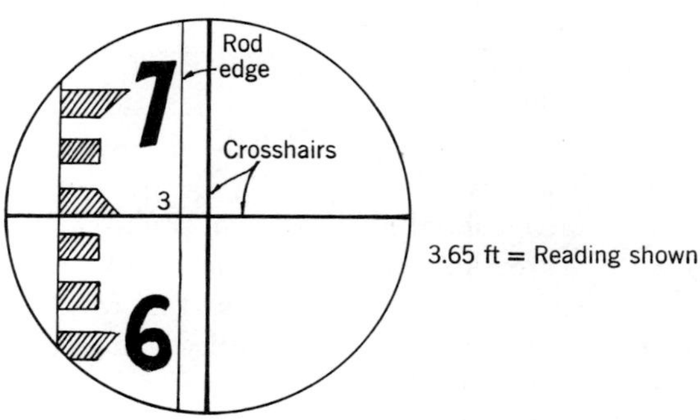

Figure 31. Philadelphia level rod.

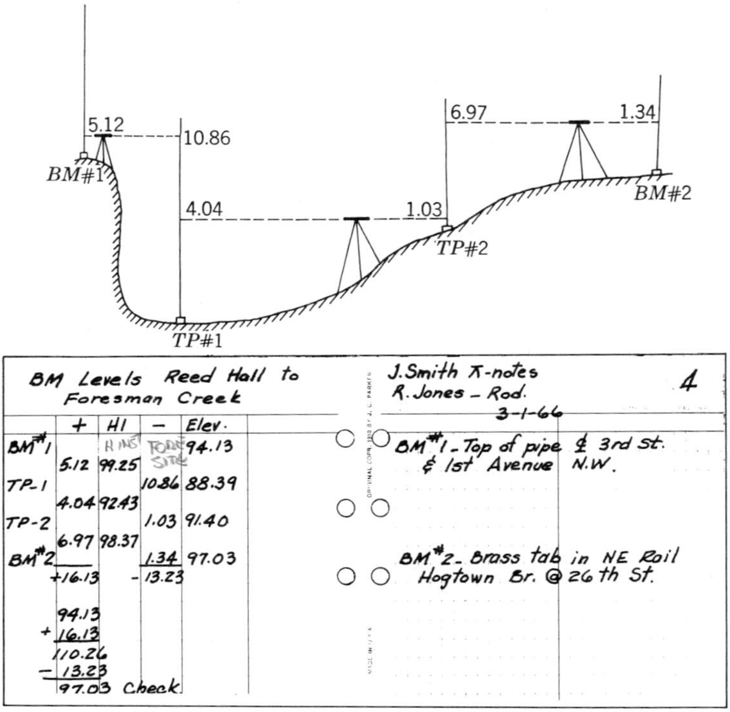

Figure 32. Differential leveling.

ment. The elevation of the instrument is the *height of instrument,* abbreviated as HI.

From the height of the instrument, you may determine the elevation of any point by a rod reading on the point, or a point can be set at any desired elevation. For example, if the height of instrument is 10.00 ft and you want to set a point at elevation 8.00 ft, the rod reading must be the difference, or 10.00 ft − 8.00 ft = 2.00 ft. After the first rod reading, all readings are subtracted from the height of instrument to determine the elevation of the point being read. These readings are called *minus* readings. In the conventional note system (Figure 32), these plus and minus rod readings are shown in the second and fourth columns. It is important that you learn to use columns in this way, and mark them as shown. Plus and minus shots are also called backsights and foresights, since this is the direction in which they occur in a line of levels.

The best (but not the cheapest) level rod is the engineer's level rod. Unlike ordinary measuring devices, the top and bottom of the marks

	+	HI	-	Elev.
BM#1	5.12	99.25		94.13
TP-1	4.04	92.43	10.86	88.39
TP-2	6.97	98.37	1.03	91.40
BM#2			1.34	97.03

Figure 33. Alternate method level notes.

rather than the centers are used to indicate the measurements. Each mark is 0.01 ft wide; the feet and tenths, as well as all even hundredths of feet, are at the top of marks. The long portion of the mark at the top of feet and tenths reminds the instrument man that the top indicates the exact point of reading. Odd hundredths, including readings ending in 0.05 ft, will be at the bottom of the marks. This is the long edge midway between the tenths.

Differential leveling is illustrated in Figure 32, with corresponding field notes. The intermediate points set for temporary use are *turning points.* The process of transferring elevations from one point to another is *running levels.*

SETTING UP THE LEVEL

The level is set up like the transit, except that since there is but one leveling bubble the level must be turned each way at right angles and parallel to pair of leveling screws, in order to read the bubble, leveling the foot plate in each direction. Since the level need not be set up over an exact point, the foot plate may be nearly leveled by moving the legs, reducing the time required as compared to leveling a transit. This is done by moving a leg which is at right angles to the telescope, crosswise to approximately the center of the bubble, and repeating with another leg. Each leg is pressed into the ground until the bubble is nearly centered when in line with that leg. Final leveling is then done with the leveling screws, as on the transit.

Remember that the first leveling with the bubble and base plate screws is approximate; the telescope is turned at right angles and roughly leveled, then exactly leveled in two directions at right angles to each other.

Since leveling in one direction throws the other direction slightly off, you waste time by attempting to center the bubble exactly in the preliminary adjustment. *Do not attempt to level the instrument so that the bubble will be centered when the telescope is rotated 180°*. Sometimes, to no avail, students will spend much of a field period trying to do this. If the instrument will not remain true when rotated end for end, you cannot correct it with the foot plate screws; if the instrument indicates that when the bubble is centered the vertical axis is not truly vertical, then adjustment of the crosshairs, wyes, or bubble tube screws is required.

If the leveling screws are so adjusted that the bubble remains in the same place in the tube when the instrument rotates, this indicates that the vertical axis is truly plumb, but this does not indicate that the telescope is level. Since this is the standard adjustment of the transit, sometimes engineers believe that the same method can be applied on a level. It cannot, since a transit is adjusted to rotate about a vertical axis. If the telescope is not level, it makes no difference for angle measurement. A level is adjusted so that the telescope is level, and it is unimportant whether the vertical axis is truly plumb.

READING THE ROD

In reading the rod, the *last* three operations determine the accuracy of the reading. These are listed here.

1. Check the bubble, which must be centered.
2. Read the rod.
3. Check the bubble, which must still be centered.

How do you know when you have completed the *last* operations? If the bubble is not centered before or after you read the rod, or if you have touched the instrument either to adjust it or accidentally, even with an eyebrow, you must repeat these three operations. This is not foolproof; of you are set up on a structure or on soft ground, the telescope may be level when you reach to one side to look at the bubble, and be off level as you shift your weight to look through the telescope. This problem is avoided in two types of instruments: those with a mirror so that the bubble may be sighted at the same time as the target, and those self-leveling instruments that do not require any check of a bubble.

The bubble must be centered exactly when the final reading is made. It must therefore be adjusted for each reading, and the bubble must be read at center at least twice for each reading.

Students often ask: What about the adjustment at right angles to the telescope? Isn't the horizontal hair sloping if the instrument is not perfectly true crosswise? This is true. The crosshair is not level, but this error is insignificant. A gross error in the bubble adjustment would be 0.1 ft in 100 ft; if the vertical hair were off the rod 1.0 ft (which it should not be), the slope of the horizontal hair would be about 0.1 in. of 1.0 ft, or 0.001 ft. This error does not increase with the distance from the target, and would be the same at 100 ft as at 400 ft. The errors that usually occur are even less than this.

• CAUSES OF ERRORS

Some common errors and mistakes in level work are:

1. Bubble not centered. The instrument man must not change his position during the final reading. This means that his feet should remain in

Figure 34. This instrument man is using the tripod as a crutch, not as a tool. *Do not touch the instrument while reading the rod.*

Figure 35. This is the proper way of reading the level bubble, without shifting the feet.

exactly the same position, no closer to the legs of the instrument than necessary, and that no part of his body or clothing should touch the instrument.

2. Faulty focusing. If either the crosshair or the objective is out of focus, a poor reading will result. The crosshair should first be focused when the objective lens is out of focus, so that the eye is not distracted.

3. Interchange of plus and minus shots entering the readings in the wrong columns. This is due to an inadequate understanding of the direction of rod readings when a student first starts field work.

4. Reading the rod a foot in error. This is common, and occurs even with experienced men. When the reading is just below an even foot, such as at 4.90 ft, the 5-ft mark is visible and the 4-ft mark is not; therefore, it is sometimes read as 5.90 ft.

5. Arithmetical errors. Students who are not obliged to do arithmetic daily are usually very poor at it. A double check of addition and subtraction is always necessary; in Figure 32 this is illustrated by adding up all plus and minus shots, and adding them and subtracting them together as an overall check on the arithmetic. Failure to do this is carelessness. Remember that you may make two errors of 1 ft each, in reading or in arithmetic, and have an even chance that they will compensate and not show up in the work at the end.

6. Reading the back of the rod. The Philadelphia pattern rod, and many others, are graduated on the back for the use of a target above 7 ft, by extending the rod so that the rodman may read the back. This requires readings on the back which are upside down, but otherwise just like the front. If a rodman holds the wrong side of the rod to an inexperienced instrument man, a reading as much as 7 ft in error may result.

FIELD EXERCISE NUMBER 4

Each student is to check out either a level or a transit and determine the elevation of one bench mark from another, using at least three intermediate turning points. It is suggested that the rods be placed by attachment to trees, buildings, special brackets, or placed by attaching plastic or metal rod faces to buildings or fences to provide turning points. In this way, no rodmen are needed, and the instructor can closely supervise the operation. Distances may be quite short, under 100 ft, in order to make the first readings easy to obtain, and to keep the students within speaking range of the instructor.

The student must make a new setup for each pair of readings; with short sights of this type, rods are often in a position where they may be bypassed, and several read from one location of the instrument. The object of the exercise is to accustom the student to the mechanics of setting up and properly reading the instrument before a long line of levels is attempted.

Two complete circuits should be completed, with the error of closing the second circuit no greater than 0.02 ft. Each student should exchange instruments between circuits, in order to become accustomed to different types of instruments. This exercise should be readily completed in three hours; many beginners will do it in two. The last four turns should take no more than twenty minutes.

Chapter 5

BENCH MARK LEVELS

The construction engineer is seldom required to run bench levels for any considerable distance, but on large construction projects there is some work of this type. The engineering surveyor must set bench mark levels before any other work can be done. Some of these level lines are hundreds of miles long.

With the engineer's level, levels may be run to an accuracy of 0.02 ft per mile, but 0.04 ft per mile is usually considered adequate, and some authorities consider 0.10 ft per mile as good as can be expected. A reasonable rate of progress is about four miles per day.

The probable error varies with the square root of the number of miles, since it is expected that errors will to some extent be made in opposite directions. This square root rule is in accordance with the theory of probability, which assumes that errors are not cumulative but are accidental; that is, they are as likely to occur in one direction as another. For various reasons, this is never entirely true, so the rule applies more accurately to shorter distances. An average error of 0.01 ft per 600-ft turn represents about 0.033 ft per ten turns, according to this rule, or about the same error per mile. Students have often been able to do this, even with builder's instruments or transits.

CONTROL ELEVATIONS

Control surveys are those made to a higher degree of precision than are ordinary mapping or preliminary surveys. That is, the control survey for mapping would be to a greater degree of precision than the mapping work itself, and would be used as a check.

The bench marks set by the U. S. Coast and Geodetic Survey are set to thousandths of a foot and are used as controls where available. Other marks are set by the State Road Departments for their own construction work and for drainage systems, and by cities to control the grade of street, sewer, and sidewalk construction. Counties often have their own system of bench marks for the same purposes. These points are usually, but not always, set to correspond with the mean sea level standard used by the U. S. C. G. S.; some large cities have their own datums.

The U. S. C. G. S. points are set for mapping purposes; the points set by other agencies are usually for some specific construction and are set as needed. Engineers for public agencies, supervising contracts for roads and drainage channels, provide bench marks at frequent intervals, and usually provide grade stakes for the contractor as well.

On building construction jobs, even though large areas of site work may be involved, the contractor does his own control survey as well as actual layout and grade stakes. Owner-builders, of course, must always do their own work or sublet it to a surveyor.

Bench marks should be provided so that it is not necessary to use a turning point in order to set a grade at any location on the project. On long narrow work, like roads, bench marks are normally set every 1000 ft. Since an instrument may be set up 300 ft from a bench mark and read 300 ft further, each point on the job may be as far as 600 ft from the nearest bench, so that this spacing is adequate. For building sites with the narrower dimension less than 1000 ft, bench marks may be set around the edge of the site; for larger jobs, they are needed within the area of the work.

It is important that all such points which are to be used for later reference be checked by closure as a *level net;* that is, that each of them be used for a turning point rather than merely determined by a single rod reading. In this way, any error will be detected in the final closure. The setting of bench marks with a single reading, without moving the instrument, is a *side shot;* that is, it is not on the main line of the level system. Grade marks are usually set this way, but there is always a possibility of error, since an error in reading the rod does not show up in the final closure.

Controls for mapping are not set so closely together; their spacing depends on the accuracy desired on the resulting map.

LEVEL RODS

Although the Philadelphia pattern rod is the most common, there is a wide variety of other rods used. Sectional rods, for folding and packing in an automobile, are available in various patterns, and some rods are graduated in feet and inches.

One type of rod is equipped with a tape which may be reeled from either end; this tape serves as the face. With this rod, the tape may be set to read elevations directly. The ordinary Philadelphia rod may be used upside down, to read elevations of floors, bottoms of beams, etc., which may be above the instrument. In such a case, the rod reading should be marked + to distinguish it from other readings, but entered in the minus column in the notes.

Rods are furnished with detachable or fixed metal faces. The detachable face is preferred, since when worn out only the face may be replaced. Rods may cost about $40 each, and contractors often use various substitutes, such as a plastic rod face, which may be bought in the form of plastic tape and then fastened to a piece of wood on the job.

For bench mark levels, a target on the rod provides greater accuracy. The rodman is given instructions to lower or raise the target until it is on the crosshair, and the rodman himself reads the rod. He may signal the reading to the levelman, or record it himself. A marker on the back of the target indicates the reading, with a vernier reading to thousandths of a foot. This same technique may be used for less accurate work when it is necessary to make shots further than the rod may be read. Lacking a target, a piece of paper may be attached to the rod, with a rubber band used as a substitute.

READING THE ROD

It is very important that the rod be plumb both ways; if it is a few inches out of plumb the reading will be in error. Normally, the rodman stands behind the rod, balancing it with his fingers; the rod will balance, if there is no wind, in substantially a plumb position (Figure 36). To plumb the rod toward the levelman, the rodman rocks it slightly on the point, forward and back; this is *waving the rod*. The levelman reads the lowest point on the rod that the crosshair reaches during this movement; this is the true reading.

Many rodmen wave the rod too far. Under ordinary circumstances, 2 or 3 in. should be sufficient; it is necessary only to wave it far enough to be sure of passing the point at which it is plumb. Also, there is a tendency to wave it too fast; the levelman needs several seconds to determine the correct reading, and it is better to wave it slowly than too rapidly.

When reading the rod to hundredths, the hair often appears to split the mark. It is theoretically immaterial which point is read in this case, but it is pointless to read the thousandth. Some authorities suggest reading the even (rather than odd) hundredth for consistency; since the tendency is always for the reading to be too great, because the rod is never exactly vertical, the lower mark may also be consistently read for slightly better results.

Figure 36*a*. Right way. In order to hold the rod plumb, it must be held at the center of the body.

The rod may be kept plumb either by attaching or holding a small level against it. These levels are either bullseye (round) or tube types, made for this purpose. They improve accuracy but are easily lost, and they are additional items for already overburdened rodmen to carry.

The usual minimum focus for leveling instruments is about 5 ft; closer readings will be blurred. In wye-type instruments, the rod may be read by sighting down the clamps at the side of the telescope, and in all instruments a reading may be made by reversing the telescope and looking into

Figure 36*b*. Wrong way. In order to hold the rod plumb, it must be held at the center of the body.

the objective end. The hairs are not visible when the telescope is used in this way, and the rod must be read by judging the center of the field. If the rod is to be read by either of these methods, it should be within two feet of the instrument. A reading may be taken also by measuring directly to the center of the eyepiece with the rod. These methods are especially useful when setting up nearly over a stake, which is used as a bench mark.

• RECIPROCAL LEVELING

In order to minimize both instrument errors and curvature of the earth, plus and minus sights are equalized. Instrument error is proportional to distance, and may be equalized by making setups so that the total foresight and total backsight distances are nearly equal.

When a bay, valley, or other obstruction must be crossed, it may be necessary, or time-saving, to take a single sight across when it is not possible to take a similar sight in the opposite direction. Such a situation is illustrated in Figure 37. In this example, the rods are read from each side of the obstruction, and the average difference is used. The errors made from one side, therefore, compensate for the errors made from the other.

In this way, level shots may be taken for as far as vision is clear and signals can be understood—as much as a mile. Obviously a target of some size must be used, with signals to indicate whether the target is to be raised or lowered. In rough country, levels could be run from one mountain top to another, although this would not be sufficiently accurate for bench marks.

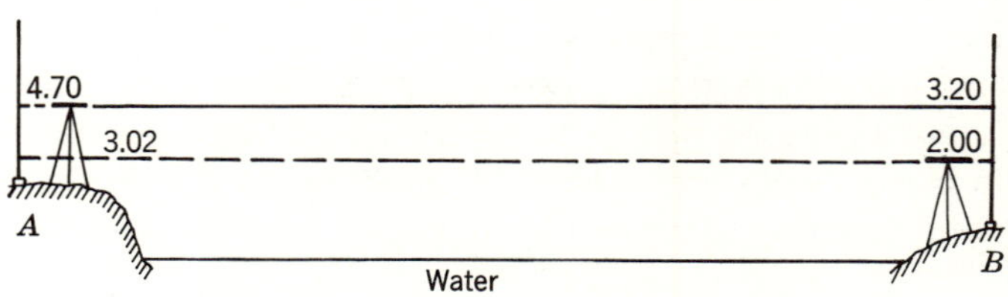

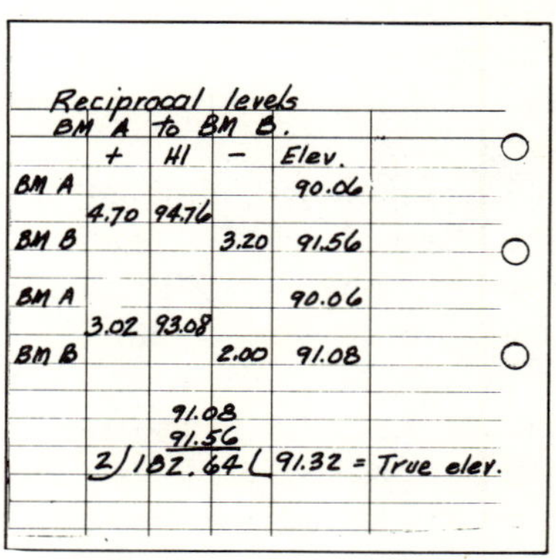

Reciprocal levels
BM A to BM B.

	+	HI	−	Elev.	
BM A				90.06	
	4.70	94.76			
BM B			3.20	91.56	
BM A				90.06	
	3.02	93.08			
BM B			2.00	91.08	
		91.08			
		91.56			
	2)182.64(			91.32 =	True elev.

Figure 37. Reciprocal Levels

LEVEL NET

In accurate work, levels are run from several directions, and each bench mark elevation determined from several lines. These lines are often of different lengths, and give different results. The elevations of all benches on the lines are adjusted by distributing the error in accordance with length of lines or number of turns.

It is probable that the error of any point on a line of levels is no greater than the error at the end. Many instances occur, however, where the contrary is true; you may run a line to a new bench mark and return to its origin with a small error, only later to find that the bench set is in error by an amount greater than the error of closure. The errors have been compensating.

FIELD NOTES

All bench marks used in a field book should be described in the same book. A simple way is to put the descriptions together in the back of the book so that it is not necessary to repeat them in the body of the notes each time the bench is used.

When books are changed during work, and when bench marks are determined by lines run in two different books, each book should be cross-referenced to the others. Numbers should be given bound field books as soon as the books are received; large firms do this in the office before books are issued so that no two books can have the same number.

DOUBLE LINES

For increased accuracy, bench mark levels are often duplicated at the time they are run. This may be done by using two turns rather than one, and keeping a double set of notes; or by using two instruments reading the same turning point.

Using two instruments is preferred, although it requires an extra man, as an independent check is made of each reading. When double turning points are used, there is a strong possibility that a foot error in reading one rod may be repeated on the second rod.

The instrument man may ride the car or truck, driven by the note-keeper. If an umbrella is used, this is also carried in the truck. If the turning points are clearly marked, back and forward rodmen may be used. In this case, the forward rodman moves up with the instrument; then it is not necessary for the crew to wait for the rodman to walk from one turning point to another.

TURNING POINTS

Turning points are not usually described in the notes; the only reason for doing so is to be able exactly to retrace a line of errors. This procedure is recommended for students, who are quite likely to make errors. Professionals prefer to use different turning points when retracing or checking a line, so that they will be less likely to repeat the same mistake at the same point.

On a paved road, a square marked on the pavement with yellow *keel* (lumber crayon) is customary, although if the rod is rocked on a flat surface, it rocks first on one edge and then on another, slightly changing the elevation. This is usually insignificant, and bench marks are made slightly rounded on top to avoid this error. A stake carried by the rodman usually makes it possible for him to install the turn at the proper distance.

When a definite distance is being used between the setup and the turning point (300 ft is typical), each man paces his distance. That is, the levelman paces from the turn to the setup, and the rodman paces from the instrument to the turn. For greater accuracy, all three hairs in the level may be read. The center hair gives the level reading, and the upper and lower hairs both check the center reading and give the distance (as is explained in Chapter 6). However, the ordinary engineer's level is not provided with these extra hairs, as they may result accidentally in erroneous readings.

HAND SIGNALS

Some of the conventional hand signals are shown in Figure 38. Most of these apply both to level and transit work, with slightly different meanings.

1. *Hold steady* or *all right* indicates that the previous action was correct. If you have been moving onto line, it indicates you are now on line; if you have been plumbing the rod, it indicates it is now plumb, or if you have been moving around to find a turning point or otherwise looking for a spot where the instrument man can see the rod, he can now see it.
2. *Move right or left* signals refer to instructions to the rodman to get on line, when running line. This signal may also mean that the rod must be moved in a certain direction to be visible in the instrument.
3. *Give me a back sight* is used when the instrument man is turning a line from the old point, as in stadia work or traverses, or in requesting a backsight in leveling. Since the rodman would normally not be at the old point except for the purpose of giving a sight, there is not likely to be

Figure 38. Hand signals. (U.S. Army)

confusion as to what point is intended; the OK (all right) signal may also be used to indicate to the rodman that his action is correct.

4. *Give me a line* is used by the head chainman when setting a point to a distance he has measured, to indicate to the instrument man to sight on a monument he has found, or as later explained, it is used in double-centering. To distinguish between "give me a line" and "this is line," the rodman may pass his hand up and down the rod to indicate *"this is line."*

5. *Plumb rod* signal is used either to set a point or to plumb a level rod. This signal is given very slowly, to distinguish it from Number 8, "wave rod," which is an almost identical motion. The arm in the plumb rod signal is held in the same direction as when the rod is off plumb, but greatly exaggerated.

6. *Establish turning point* is a signal that the levelman can no longer see the level rod—either because of elevation, interferences, or distance—during leveling work other than bench mark levels, and that he must move up. When running bench mark levels, the rodman knows when the next turning point is needed.

7. *This is a turning point* is the same as the number 4 signal, but with the level rod. It also distinguishes an exact rod reading from an approximate one. The signal may be given either in response to the levelman's order, or because the rodman foresees the necessity of a turning point.

8. *Wave the rod* means to slowly wave the rod toward and away from the levelman. If the rodman is close enough to distinguish the signal clearly, the levelman waves toward the rodman rather than to the side.

9. *Face the rod* is a signal which really means "get on the job," because the rodman should face the level rod to the levelman in any case. It indicates the rodman has been careless in holding the rod, or has been waiting for the instrument man to set up or complete some other work.

10. *Reverse the rod* reminds the rodman that he holds the rod with the wrong face to the levelman.

11. *Boost the rod* means the rod is below the line of sight of the levelman, and the level rod is to be extended to its full length. If the rod is already extended, it means the rod must be lifted still further; for some work, when the shots are approximate, the rodman measures 1 to 3 ft on his body, and "boosts" the rod this distance. The rodman calls to the notekeeper "boosting 3 ft" or whatever it may be, and the notekeeper adds this to the rod reading. This appears in the notes as a rod reading in excess of the length of the rod.

12. *Move forward, move back* may be used under a variety of circumstances, including guiding a rodman to where he should take ground shots.

13. *Up* and *down* may be used under a variety of conditions, particu-

larly when giving instructions to a rodman who is raising or lowering a target on the rod.

14. *Pick up the instrument* deserves a special signal, because accidentally picking up the instrument when not intended (as giving a signal to another man, or signaling someone to return to the truck) may cause observations to be repeated. It is a move forward signal, especially for the instrument man, and indicates that work at the forward position, for which line from the transit was required, has been completed.

15. *Come in* may refer to one rodman, or it may be a general "quit work" signal.

Of the signals not illustrated, the *double center* is important. It is used for extending the line accurately, and consists of rotating the range rod, similar to the "give me a line" signal, but the rod is rotated through the full 360°. The *can't see* signal is to wave a hand or flag back and forth below the waist, and assumes that the rodman knows what to do to get into the line of sight.

At a distance over a quarter mile, hands are not readily seen, and at half-mile or longer distances only a flag can be seen by the naked eye. The instrument man then can give a limited number of signals—the flag on one side or the other to indicate the direction in which the rodman should move, back and forth over the head to indicate OK, or under the waist for *can't see,* and a full circle for "come home," the equivalent of the *come in* signal indicated.

The instrument man can see the rodman's signals readily at one or two miles. The rodman, therefore, must direct the crew's operations under these conditions. It is important that a rodman learn the signals his crew uses as soon as possible, because he cannot be effective as a worker if he must be given specific directions orally, The head chainman particularly must be able to indicate unexpected conditions to the instrument man, and in many cases must decide what he will direct the instrument man to do.

READING IN THOUSANDTHS

The ordinary level rod may be readily read to thousandths at the distances encountered in building construction work, and there is no objection to doing so, if the levelman feels the additional accuracy is useful. However, such reading must be consistent; if *any* readings are to be made in this way, all of them must be. This is necessary to assure that hundredths and thousandths will not be confused in addition and subtraction. In addition, as pointed out in the section on significant figures, no accuracy is gained by adding numbers when one of them is in smaller units than the others. Usually a target is used and readings are made by the rodman.

• MISTAKES AND ERRORS

Here are some of the common sources of error on bench mark leveling.

1. *Instrument not in adjustment.* The method of adjustment is described later in this text; the instrument should be checked regularly. When running bench levels, the check should be daily. Marks may be set at the office for this purpose, so the check may be made in a very few minutes. This check should not hold up the crew; the levelman need only finish breakfast a few minutes early and run the check before the crew is ready to leave. On construction jobs, a weekly check is sufficient, but closures should be made during the week so that an error in the instrument will be discovered.

2. *Temperature variations in the instrument.* This is avoided by covering the instrument with an umbrella. For single sights and ordinary construction work, these errors are not large, but for bench mark leveling they may become significant.

3. *Wind.* It is very difficult to keep an instrument steady on windy days. It is preferable not to attempt precise work at such times.

4. *Settlement of the instrument between shots* may take place in soft ground, and is especially noticeable on hot pavement. Setups should be on ground in preference to asphaltic pavement.

5. *Movement of the instrument during reading* is noticeable in soft ground where the instrument not only moves as the levelman moves to read the bubble, but moves back when he reads the instrument. It is especially important that the levelman shift his weight as little as possible during reading; even though the feet are held in the same place, some shifting is necessary. It may be necessary to have a second man check the level during the shot. The use of the type of level that allows the bubble to be seen from the rear is highly recommended.

In construction work, it is often necessary to set up on formwork or on structural steel. At these locations, careful attention to supporting structure will minimize movement of the instrument.

6. *Variation of the turning point* may occur because of carelessness in driving a turning point stake hard enough, clearly marking a point, or from choosing an improper point. The most common student error (which is a mark of the amateur) is to "assume" a point and "assume" that one can find it again without a mark; usually, because the student forgot a marking crayon. A standing joke among rodmen is to take a turn on one's

toe or the bumper of a car. By moving the turning point, it is then unnecessary to run the intervening line!

7. *Improperly raised rod.* Sometimes the rod will stick before reaching full length. As a result, the rod is short and an error results if the next reading is on the lower part of the rod.

8. *Improper recording.* If a levelman is reading dimensions to the notekeeper, they should always be read in the same way, preferably in a military manner. For example, never read "Five-seven," "Five-seventy," or even "Five-oh-seven," for "Five-point-zero-seven."

9. *Anticipating readings.* The levelman should never attempt to figure what his reading should be, before closing on a bench. Likewise, the notekeeper should not compute the closing reading, but check it after it has been recorded. The tendency to read the expected figure, especially the foot, is overwhelming.

FIELD EXERCISE NUMBER 5

Each two-man crew will draw an instrument, level rod, and appropriate turning point material, and run a line of bench mark levels from a given mark to another a half-mile away, and return.

Expected accuracy each way is 0.02 ft to 0.04 ft, depending on the type of instrument used. Record both the number and type of instrument. Each student runs the line one way, and his partner uses this elevation to return the line.

On slopes, the ground usually appears much more nearly level than it really is, and beginning surveyors frequently waste a great deal of their time making setups and setting turning points that are too high or low to be used. You will save time by checking all setups and turns first by rough leveling. This is done on the setup by balancing the instrument with legs together, and pointing it so the bubble is approximately level, looking down the side of the telescope; if you cannot see the rod, then you know this is not the place for a setup. When the rodman is setting a turning point, the instrument man watches him in the telescope. The instrument man can tell either by a rod reading or by looking at the rodman's feet, depending on whether the rodman is going uphill or downhill, when a useful turning point is reached.

Chapter 6

THE STADIA METHOD

STADIA PRINCIPLES

Stadia method refers, strictly speaking, to measurement of distance using the apparent angle formed by the ends of a rod of known length. An optical range-finder measures in this way, and an artilleryman is trained to measure distance by the apparent size of an object as viewed between his knuckles at arm's length. Transits are equipped with crosshairs at such a distance apart that on a rod 100 ft away, the hairs appear to be 1 ft apart. Consequently, at any distance, the measurement between the crosshairs on a vertical rod may be multiplied by 100 to find the distance to the rod.

Determining distance by this method is must faster, but less accurate, than chaining. Since the level rod may normally be read to the nearest 0.01 ft, the error in the resulting distance will be about 1 ft. Mapping is frequently done by stadia measurements, getting elevations by measuring vertical angles at the same time as locations are obtained by horizontal angles and distances by stadia. This process is referred to as *mapping by stadia,* and *stadia* is often used to refer to measurement of both elevations and distances. It is not necessary, however, that vertical angles be used to determine elevations. If the elevation of a point is no more than 5 ft above or below the ground at the instrument, a level reading may be taken for determining the elevation.

Stadia mapping is most efficient for rough areas. To make maps of such areas, you must determine the ground elevation only at controlling points —the bottoms of streams, tops of hills, tops of ridges, and the like—and to determine the locations at which these elevations are obtained. This information is plotted on the map, and from these points the elevation of the rest of the surface is estimated. Figure 39 shows how one such elevation, or shot, is obtained. The instrument is set up at A, from which the instrument man wants to obtain the elevation and location of the ground at B. He has the elevation of the instrument, from reading the rod as a level

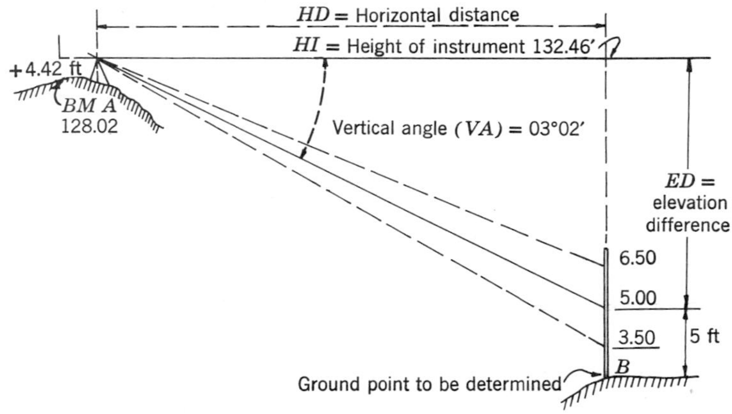

Exercise No. 6 — Stadia Practice (127.5)
H.I.= 4.42+128.02= 132.46

J. Smith ⊼ notes — 5
3-26-66

	U.H.	CH	LH	Azim.	V.A.	Dist.	HD/100	HD	ED/100	ED	ELEV.
BM A		4.42			L					4.42	128.02
	6.50	5.00	3.50	42-07	-03-02	300	99.72	299	-5.28	-15.9	111.6
	5.72	4.67	3.62	175-00	L	210	—	210	—	-4.7	127.8

Figure 39. Principles of stadia.

on the bench mark shown; and he knows the direction of a base line from point *A*, so that he can determine the directions of other stakes.

The rodman places the bottom of the rod on the ground at *B*, and the instrument man sets the center horizontal hair on the 5-ft mark on the rod. He reads the upper and lower hairs, here 3.50 ft and 6.50 ft, giving him a difference of 3.00 ft. This indicates a distance of 300 ft to the rod. This distance is not quite correct, for two reasons; he has measured it on a slope, while he wants the horizontal distance; and he is looking downward with the rod rather than directly level with it. These corrections are combined in stadia tables.

The instrument man reads his vertical angle on the transit, and for this vertical angle he finds the fall per 100 ft in the line; from this he figures the difference in elevation between the instrument and the 5-ft mark on the rod.

ILLUSTRATIVE EXAMPLE

The diagram and notes in Figure 39 show two stadia shots. The instrument man has set up at *A*, in sight of the bench mark *A* of elevation 128.02

ft. He reads the level rod on this mark, obtaining +4.42 ft as his distance above it. This is entered as a center hair reading of 4.42 ft in the notes, with *L* indicating that the line was level; that is, there was no vertical angle. The *ED* (elevation difference) is therefore 4.42 ft. Above, on the left side of the note page, he shows the addition 128.02 ft + 4.42 ft = 132.46 ft, the height of the instrument. He subtracts 5 ft from this elevation, obtaining 127.5 ft, which is marked just above, and will be used later. He now reads the vertical angle to point *B* of −03−02, which the stadia hair intercepts on the rod to determine distance, and the azimuth to determine direction. The figures on the left half of the field note page are field data; those on the right side are calculations that are normally done in the office. The 5-ft mark on the rod is used as a convenience, a target being attached to the rod at this mark. This figure is easily subtracted and added, and the target is high enough to be free of underbrush. Other marks can be, and often are, used, especially in woods where it is hard to see the rod.

Only two crosshair readings are necessary, since the distance from either upper or lower crosshair to the center crosshair is the same. A half reading can be read and then doubled for the full reading. The difference between the upper and lower hair readings multiplied by 100 is the approximate slope distance, to which corrections are made to determine the horizontal true distance. The approximate slope distance is the distance used to determine the elevation difference.

The distance having been determined as 300 ft, the tables shown in Figure 40 are consulted for the vertical angle of −03−02. (Notice that the sign is always used on both the vertical angle and on elevation differences.) *Dist* (distance) column gives 99.72° corresponding to this angle. This means that for a rod reading of 1.00 ft, corresponding to a distance of 100 ft, the true horizontal distance is 99.72 ft. The *correction* per 100 ft is, therefore, the difference between 99.72 and 100.00 ft, or 0.28 ft. It is easier to multiply by .28 than by 99.72 ft. For 300 ft, the correction is 0.84 ft, or 3 ft × .28 ft; since you are figuring to the nearest foot, call this 1 ft. The true horizontal distance is then 299 ft, shown in the notes under *HD* (horizontal distance).

To determine the elevation of the ground point, the same table is used to find the difference in elevation per 100 ft (*ED*/100 in the notes), labeled *Elev* in the table. For a vertical angle of −03−02, this difference in elevation is −5.28 ft per 100 ft. This is multiplied by 3 for a difference in 300 ft of 15.9 ft, to the nearest 0.1 ft. Note that to be sure of the nearest 0.1 ft in 15.9 ft, we must have 5.28 ft to 0.01 ft; three places in one multiplier give three in the answer. One must have hundredths in the table to get tenths in the answer.

Subtracting 15.9 ft from the *HI*, you get 132.46 ft − 15.9 ft = 116.6 ft, which is the elevation of the rod point that the transit horizontal center

Stadia Reductions for Rod Readings of 100

Minutes	10°		11°		12°		13°		14°	
	Hor. Dist.	Diff. Elev.	Hor. Dist.	Diff. Elev.	Hor. Dist.	Diff. Elev.	Hor. Dist.	Diff. Elev.	Hor. Dist.	Diff. Elev.
0′	96.98	17.10	96.36	18.73	95.68	20.34	94.94	21.92	94.15	23.47
2	96.96	17.16	96.34	18.78	95.65	20.39	94.91	21.97	94.12	23.52
4	96.94	17.21	96.32	18.84	95.63	20.44	94.89	22.02	94.09	23.58
6	96.92	17.26	96.29	18.89	95.61	20.50	94.86	22.08	94.07	23.63
8	96.90	17.32	96.27	18.95	95.58	20.55	94.84	22.13	94.04	23.68
10	96.88	17.37	96.25	19.00	95.56	20.60	94.81	22.18	94.01	23.73
12	96.86	17.43	96.23	19.05	95.53	20.66	94.79	22.23	93.98	23.78
14	96.84	17.48	96.21	19.11	95.51	20.71	94.76	22.28	93.95	23.83
16	96.82	17.54	96.18	19.16	95.49	20.76	94.73	22.34	93.93	23.88
18	96.80	17.59	96.16	19.21	95.46	20.81	94.71	22.39	93.90	23.93
20	96.78	17.65	96.14	19.27	95.44	20.87	94.68	22.44	93.87	23.99
22	96.76	17.70	96.12	19.32	95.41	20.92	94.66	22.49	93.84	24.04
24	96.74	17.76	96.09	19.38	95.39	20.97	94.63	22.54	93.81	24.09
26	96.72	17.81	96.07	19.43	95.36	21.03	94.60	22.60	93.79	24.14
28	96.70	17.86	96.05	19.48	95.34	21.08	94.58	22.65	93.76	24.19
30	96.68	17.92	96.03	19.54	95.32	21.13	94.55	22.70	93.73	24.24
32	96.66	17.97	96.00	19.59	95.29	21.18	94.52	22.75	93.70	24.29
34	96.64	18.03	95.98	19.64	95.27	21.24	94.50	22.80	93.67	24.34
36	96.62	18.08	95.96	19.70	95.24	21.29	94.47	22.85	93.65	24.39
38	96.60	18.14	95.93	19.75	95.22	21.34	94.44	22.91	93.62	24.44
40	96.57	18.19	95.91	19.80	95.19	21.39	94.42	22.96	93.59	24.49
42	96.55	18.24	95.89	19.86	95.17	21.45	94.39	23.01	93.56	24.55
44	96.53	18.30	95.86	19.91	95.14	21.50	94.36	23.06	93.53	24.60
46	96.51	18.35	95.84	19.96	95.12	21.55	94.34	23.11	93.50	24.65
48	96.49	18.41	95.82	20.02	95.09	21.60	94.31	23.16	93.47	24.70
50	96.47	18.46	95.79	20.07	95.07	21.66	94.28	23.22	93.45	24.75
52	96.45	18.51	95.77	20.12	95.04	21.71	94.26	23.27	93.42	24.80
54	96.42	18.57	95.75	20.18	95.02	21.76	94.23	23.32	93.39	24 85
56	96.40	18.62	95.72	20.23	94.99	21.81	94.20	23.37	93.36	24.90
58	96.38	18.68	95.70	20.28	94.97	21.87	94.17	23.42	93.33	24.95
$c+f=.75$	.74	.14	.73	.15	.73	.16	.73	.17	.73	.19
$c+f=1.00$	.98	.18	.98	.20	.98	.22	.97	.23	.97	.25
$c+f=1.25$	1.23	.23	1.22	.25	1.22	.27	1.21	.29	1.21	.31
Minutes	**15°**		**16°**		**17°**		**18°**		**19°**	
0′	93.30	25.00	92.40	26.50	91.45	27.96	90.45	29.39	89.40	30.78
2	93.27	25.05	92.37	26.55	91.42	28.01	90.42	29.44	89.36	30.83
4	93.24	25.10	92.34	26.59	91.39	28.06	90.38	29.48	89.33	30.87
6	93.21	25.15	92.31	26.64	91.35	28.10	90.35	29.53	89.29	30.92
8	93.18	25.20	92.28	26.69	91.32	28.15	90.31	29.58	89.26	30.97
10	93.16	25.25	92.25	26.74	91.29	28.20	90.28	29.62	89.22	31.01
12	93.13	25.30	92.22	26.79	91.26	28.25	90.24	29.67	89.18	31.06
14	93.10	25.35	92.19	26.84	91.22	28.30	90.21	29.72	89.15	31.10
16	93.07	25.40	92.15	26.89	91.19	28.34	90.18	29.76	89.11	31.15
18	93.04	25.45	92.12	26.94	91.16	28.39	90.14	28.81	89.08	31.19
20	93.01	25 50	92.09	26.99	91.12	28.44	90.11	29.86	89.04	31.24
22	92.98	25.55	92.06	27.04	91.09	28.49	90.07	29.90	89.00	31.28
24	92.95	25.60	92.03	27.09	91.06	28.54	90.04	29.95	88.96	31.33
26	92.92	25.65	92.00	27.13	91.02	28.58	90.00	30.00	88.93	31.38
28	92.89	25.70	91.97	27.18	90.99	28.63	89.97	30.04	88.89	31.42
30	92.86	25.75	91.93	27.23	90.96	28.68	89.93	30.09	88.86	31.47
32	92 83	25.80	91.90	27.28	90.92	28.73	89.90	30.14	88.82	31.51
34	92.80	25.85	91.87	27.33	90.89	28.77	89.86	30.19	88.78	31.56
36	92.77	25.90	91.84	27.38	90.86	28.82	89.83	30.23	88.75	31.60
38	92.74	25.95	91.81	27.43	90.82	28.87	89.79	30.28	88.71	31.65
40	92.71	26.00	91.77	27.48	90.79	28.92	89.76	30 32	88.67	31 69
42	92.68	26.05	91.74	27.52	90.76	28.96	89.72	30.37	88.64	31.74
44	92.65	26.10	91.71	27.57	90.72	29.01	89.69	30.41	88.60	31.78
46	92.62	26.15	91.68	27.62	90.69	29.06	89.65	30.46	88.56	31.83
48	92.59	26.20	91.65	27.67	90.66	29.11	89.61	30.51	88.53	31.87
50	92.56	26.25	91.61	27.72	90.62	29.15	89.58	30.55	88.49	31.92
52	92.53	26.30	91.58	27.77	90.59	29.20	89.54	30.60	88.45	31.96
54	92.49	26.35	91.55	27.81	90.55	29.25	89.51	30.65	88.41	32.01
56	92.46	26.40	91.52	27.86	90.52	29.30	89.47	30.69	88.38	32 05
58	92.43	26.45	91.48	27.91	90.48	29.34	89.44	30.74	88.34	32.09
$c+f=.75$	.72	.20	.72	.21	72	.23	.71	.24	.71	.25
$c+f=1.00$	.96	.27	.96	.28	95	.30	.95	.32	.94	.33
$c+f=1.25$	1 20	.34	1.20	.36	1.19	.38	1.19	.40	1.18	.42

Figure 40. Stadia Tables. (Courtesy of Lefax Co.)

Stadia Reduction for Rod Readings of 100

Minutes	0° Dist.	0° Elev.	1° Dist.	1° Elev.	2° Dist.	2° Elev.	3° Dist.	3° Elev.	4° Dist.	4° Elev.
0′	100.00	.00	99.97	1.74	99.88	3.49	99.73	5.23	99.51	6.96
2	100.00	.06	99.97	1.80	99.87	3.55	99.72	5 28	99.51	7.02
4	100.00	.12	99.97	1.86	99.87	3.60	99.71	5.34	99.50	7.07
6	100.00	.17	99.96	1.92	99.87	3.66	99.71	5.40	99.49	7.13
8	100.00	.23	99.96	1.98	99.86	3.72	99.70	5.46	99.48	7.19
10	100.00	.29	99.96	2.04	99.86	3.78	99.69	5.52	99.47	7.25
12	100.00	.35	99.96	2.09	99.85	3.84	99.69	5.57	99.46	7.30
14	100.00	.41	99.95	2.15	99.85	3.90	99.68	5.63	99.46	7.36
16	100.00	.47	99.95	2.21	99.84	3.95	99.68	5.69	99.45	7.42
18	100.00	.52	99.95	2.27	99.84	4.01	99.67	5.75	99.44	7.48
20	100.00	.58	99.95	2.33	99.83	4.07	99.66	5.80	99.43	7.53
22	100.00	.64	99.94	2.38	99.83	4.13	99.66	5.86	99.42	7.59
24	100.00	.70	99.94	2.44	99.82	4.18	99.65	5.92	99.41	7.65
26	99.99	.76	99.94	2.50	99.82	4.24	99.64	5.98	99.40	7.71
28	99.99	.81	99.93	2.56	99.81	4.30	99.63	6.04	99.39	7.76
30	99.99	.87	99.93	2.62	99.81	4.36	99.63	6.09	99.38	7.82
32	99.99	.93	99.93	2.67	99.80	4.42	99.62	6.15	99.38	7.88
34	99.99	.99	99.93	2.73	99 80	4.48	99 62	6.21	99.37	7.94
36	99.99	1.05	99.92	2.79	99.79	4.53	99.61	6.27	99.36	7.99
38	99.99	1.11	99.92	2.85	99.79	4.59	99.60	6.33	99.35	8.05
40	99.99	1.16	99.92	2.91	99.78	4.65	99.59	6.38	99.34	8.11
42	99.99	1.22	99.91	2.97	99.78	4.71	99.59	6.44	99.33	8.17
44	99.98	1.28	99.91	3.02	99.77	4.76	99.58	6.50	99.32	8.22
46	99.98	1.34	99.90	3.08	99.77	4.82	99.57	6.56	99.31	8.28
48	99.98	1.40	99.90	3.14	99.76	4.88	99.56	6.61	99.30	8.34
50	99.98	1.45	99.90	3.20	99.76	4.94	99.56	6.67	99.29	8.40
52	99.98	1.51	90.89	3.26	99.75	4.99	99.55	6.73	99.28	8.45
54	99.98	1.57	99.89	3.31	99.74	5.05	99.54	6.78	99.27	8.51
56	99.97	1.63	99.89	3.37	99.74	5.11	99.53	6.84	99.26	8.57
58	99.97	1.69	99 88	3.43	99.73	5.17	99.52	6.90	99.25	8.63
$c+f=.75$	.75	.01	.75	.02	.75	.03	.75	.05	.75	.06
$c+f=1.00$	1.00	.01	1.00	.03	1.00	.04	1 00	.06	1.00	.08
$c+f=1.25$	1.25	.02	1.25	.03	1.25	.05	1.25	.08	1.25	.10
Minutes	**5° Dist.**	**5° Elev.**	**6° Dist.**	**6° Elev.**	**7° Dist.**	**7° Elev.**	**8° Dist.**	**8° Elev.**	**9° Dist.**	**9° Elev.**
0′	99.24	8.68	98.91	10.40	98.51	12.10	98.06	13.78	97.55	15.45
2	99.23	8.74	98.90	10.45	98.50	12.15	98.05	13.84	97.53	15.51
4	99.22	8.80	98.88	10.51	98.48	12.21	98.03	13.89	97.52	15.56
6	99.21	8.85	98.87	10.57	98.47	12.26	98.01	13.95	97.50	15.62
8	99.20	8.91	98.86	10.62	98.46	12.32	98.00	14.01	97.48	15.67
10	99.19	8.97	98.85	10.69	98.44	12.38	97.98	14.06	97.46	15.73
12	99.18	9.03	98.83	10.74	98.43	12.43	97.97	14.12	97.44	15.78
14	99.17	9.08	98.82	10.79	98.41	12.49	97.95	14.17	97.43	15.84
16	99.16	9.14	98.81	10.85	98.40	12.55	97.93	14.23	97.41	15.89
18	99.15	9.20	98.80	10.91	98.39	12.60	97.92	14.28	97.39	15.95
20	99.14	9.25	98.78	10.96	98.37	12.66	97.90	14.34	97.37	16.00
22	99.13	9.31	98.77	11.02	98.36	12.72	97.88	14.40	97.35	16.06
24	99.11	9.37	98.76	11.08	98.34	12.77	97.87	14.45	97.33	16.11
26	99.10	9.43	98.74	11.13	98.33	12.83	97 85	14.51	97.31	16.17
28	99.09	9.48	98.73	11.19	98.31	12.88	97.83	14.56	97.29	16.22
30	99.08	9.54	98.72	11.25	98.29	12.94	97.82	14.62	97.28	16.28
32	99.07	9.60	98.71	11.30	98.28	13.00	97.80	14.67	97.26	16.33
34	99.06	9.65	98.69	11.36	98.27	13.05	97.78	14.73	97.24	16.39
36	99.05	9.71	98.68	11.42	98.25	13.11	97.76	14.79	97.22	16.44
38	99.04	9.77	98.67	11.47	98.24	13.17	97.75	14.84	97.20	16.50
40	99.03	9.83	98.65	11.53	98.22	13.22	97.73	14.90	97.18	16.55
42	99.01	9.88	98.64	11.59	98.20	13.28	97.71	14.95	97.16	16.61
44	99.00	9.94	98.63	11.64	98.19	13.33	97.69	15.01	97.14	16.66
46	98.99	10.00	98.61	11.70	98.17	13.39	97.68	15.06	97.12	16.72
48	98.98	10.05	98.60	11.76	98.16	13.45	97.66	15.12	97.10	16.77
50	98.97	10.11	98.58	11.81	98.14	13 50	97.64	15.17	97.08	16.83
52	98.96	10.17	98.57	11.87	98.13	13.56	97.62	15.23	97.06	16.88
54	98.94	10.22	98.56	11.93	98.11	13.61	97.61	15.28	97.04	16.94
56	98.93	10.28	98.54	11.98	98.10	13.67	97.59	15.34	97.02	16.99
58	98.92	10.34	98.53	12.04	98.08	13.73	97.57	15.40	97 00	17.05
$c+f=.75$	.75	.07	.75	.08	.74	.10	.74	.11	.74	.12
$c+f=1.00$	.99	.09	.99	.11	.99	.13	.99	.15	.99	.16
$c+f=1.25$	1.24	.11	1.24	.14	1.24	.16	1.23	.18	1.23	.21

Figure 40. Continued

hair is set on; that is, the elevation of the 5-ft mark on the rod. To find the ground elevation, you must subtract another 5 ft, to get 111.6 ft.

However, since this 5 ft must be subtracted from *every* shot, it is easier to subtract it from the *HI* in the first place, using 127.5 ft for a figure to subtract *from*. This is the reason for showing this figure on the notes. Subtracting, 127.5 ft − 15.9 ft = 111.6 ft, or the same ground elevation, with but one operation.

READING WITHOUT VERTICAL ANGLE

If the ground is at such an elevation that the rod can be read with the transit telescope level, both readings and calculations are easier; the next line in the notes illustrates such a shot. The telescope is leveled with the bubble, and the hairs read as before. Distance is computed as with a vertical angle, but since there is no vertical angle correction, the distance between the hairs (the *intercept* distance) multiplied by 100 will be the true distance. In the example of notes shown, *L* is used rather than 00-00 to indicate that the bubble was used. The 0° reading on the vertical circle may not represent the same setting as true level, due to an index error. The elevation difference is read directly from the center hair when using a level telescope. Notice that this elevation difference includes the 5-ft correction, so it is subtracted from the true *HI*. The two methods of computing elevation difference may be used on the same page if care is taken to remember to apply the proper deduction of the elevation difference (to allow for target setting when vertical angles are used).

READING THE VERTICAL CIRCLE

On the horizontal circle, the graduations are on the outside of the index marker; on the vertical circle, the graduated circle is on the *inside* of the index. Therefore, having learned how to read the horizontal circle is no help in reading the vertical circle; *practically every student reads the vertical circle incorrectly at first.* The horizontal circle is always read clockwise, but the vertical circle is read alternately clockwise and counterclockwise, which compounds the confusion.

A reading on the vertical circle is shown in Figure 41. The upper, or inside, zero mark has moved to the left, which indicates that the telescope, which points to the left, has moved upward; the reading is therefore +. The degree is read as just past the 1° mark, where the lower index intersects the upper circle. The number of minutes in excess of the 1° mark is read on the vernier *in the same direction* as you have read the

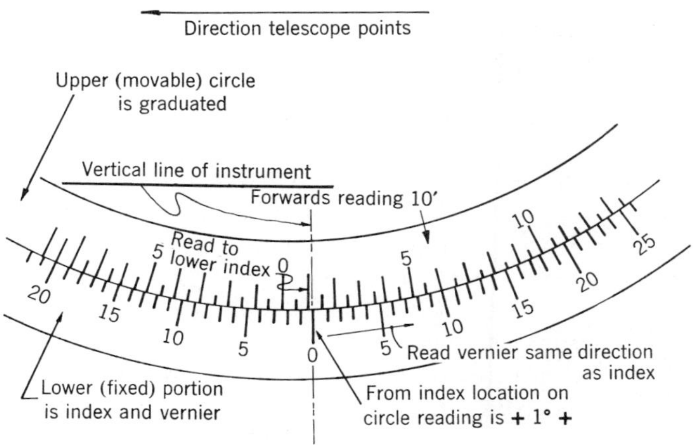

Figure 41. Reading the vertical circle.

degree on the upper (graduated) circle—to the right, in this instance. The + and − marks on the figure do not appear on the instrument, but they may be marked on the vernier in pencil.

These are the common errors made in reading the vertical circle.

1. Reading the wrong index. In this instance, if the upper index is read against the lower vernier, you have a reading between 2 and 3—which means nothing at all, since the lower graduations are not degrees.

2. Reading the corresponding lines on the wrong circle. In this instance, the corresponding line is 10′; if you read the 6° mark above, it means nothing.

3. Reading the vernier in the wrong direction. In the figure, the 20′ line on the left part of the vernier also corresponds to the degree mark, but it means you are 20′ *short* of 1°30′, not reading an angle of 1°20′. Since you use either vernier, depending on whether the vertical angle is plus or minus, you may not merely choose the left-hand vernier all the time, as you do with the horizontal circle.

4. Recording the angle as plus rather than minus, or vice versa. Put your finger on the *lower* index and think to yourself: Which way did the telescope have to move to get in such a position? If it was upwards, the reading is plus, and vice versa.

In summary, remember that *the vernier may be read either direction, according to the direction the angle is turned;* and that the *lower index is the center of the instrument.*

SIGNIFICANT FIGURES

It is necessary to the understanding of any computations that the student understand "What is a number?". Fortunately this is not really important most of the time. One can order a 2 × 4 or a 5-lb sack of sugar without worrying about what these numbers represent. One can count five cows in a field, with no argument about, "How many cows are five?"

But these numbers are not measurements of physical quantities. Everyone knows a 2 × 4 is not really 2″ × 4″, and a 5-lb bag of sugar is that because the manufacturer knew beforehand, with specied tolerances, how much sugar must go into the bag. Cows come in units—at least before they are butchered, and we are not concerned with the actual quantity of a cow. But how do we describe physical quantities—measurements of preexisting quantities, or ratios which are not even?

Take for example, pi, the ratio between the circumference and diameter of a circle. Is it 3? 3.1? 3.14? 3.142? 3.1416? All of these numbers are incorrect in the sense that the quantity can be determined more accurately; and all are correct in the sense that each number correctly represents the value to the number of digits shown. That is, the digits shown are correct, but they do not represent the same *range of values.* To know which value of pi is to be used in a certain case, one must know the precision necessary.

3 means a value between 2.5 and 3.5 to the nearest digit; that is, it represents one significant figure. Since pi falls in that range, 3 is correct. 3.0 is another quantity, however; it represents a range between 2.95 and 3.05. The same is true of each number used for pi above; the last figure is nearer than any other number.

Therefore, if I were to ask the circumference of a 3-ft circle, the correct answer is 9 ft. For a 3.0-ft circle, it would be 9.3 ft. To determine a circumference, multiplying 3 ft by 3.1416 would give 9.4248 ft, but we don't know the answer this closely. Suppose the diameter were really 2.51 ft? *One should not use figures in the result which are meaningless as to physical data; this assumes data not given, and therefore is wrong.*

When answers are computed in this way and are known to be misleading, they should be "rounded off" to the nearest figure which is known to be correct, or which is at least more likely correct than any other. 5.1 × 5.1, if multiplied by arithmetic, is 26.01. If we are using the rule of significant figures, the answer is 26. Since 5.1 may vary from 5.05 to 5.15, the actual result could be anywhere between 25.50 and 26.52.

This rule is but an approximation, as the actual variation is in percentage. That is, the accuracy of a number in terms of percentage error depends on the number—a significant figure of 5, for example, represents a possible variation of 11%. For this reason, an extra figure is sometimes retained when the first number is low. In multiplying 1.1 × 0.9, 1 × 0.9 should be used to preserve the same number of figures in each number; but it is slightly more accurate to keep 1.1 × 0.9 = 0.99; 0.99 is more nearly the probable value, which varies from 0.89 to 1.09.

When adding, like units must be added, and these may have no relationship to significant figures. For example, if I am 8, and my brother is 0.1 year older, how old is my brother? Since I may be (ignoring the contrary custom, but in terms of scientific notation) from 7.5 to 8.5 years old, there is no point in saying my brother is 8.1 years old—a totally misleading number. For the same reason, if levels are read to hundredths, there is nothing gained by reading some of them to thousandths.

Similarly, there is no direct relationship between location of a decimal point and number of significant figures. The numbers 5, 0.05, 0.005, and 0.0005 all have but one significant figure. This may also be illustrated with metric units. If I have 0.334 meters, this is 33.4 centimeters, or 334 millimeters—in each case there are still three significant figures, but expressed in different units. This occurs with other types of units, not so simply related as metric units.

How about 50 and 500? At this point, our numbering system fails—how many significant figures are there in 50? Does it represent the range between 45 and 55, or between 49.5 and 50.5? Insufficient information is given to know this. Unless one knows the figures used to obtain a number ending in 0, the actual value is not definite.

This is not true if zeros are added to the right of the decimal point, since they are not necessary except to show precision. 50.0 for example, has three significant figures, as does 0.00500. The zeros to the left of the 5 in 0.00500 are not significant figures; they are necessary to show how the figures are to be applied; that is, to show units.

In summary, numbers should always be recorded to the precision read, no more and no less. This rule is somewhat modified in the reading of minutes and seconds; it is customary to use seconds to describe portions of minutes, although angles are not usually read with this degree of precision.

There is no direct relationship, unless you know what kind of problem you have, between the number of significant figures in the data given for a problem and the number of significant figures in the answer. The number of significant figures in the answer can be determined by varying the data within the limits of the number of significant figures given. For example, if a given measurement is 1.0 ft, try figuring the problem with 0.95 and 1.05 ft to see how much variation you can expect in the answer.

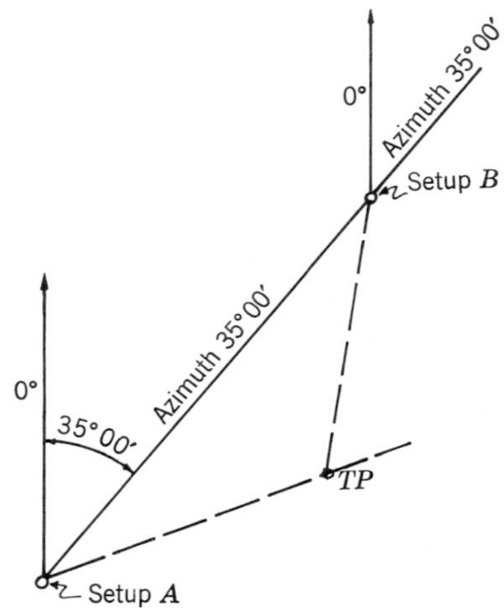

Figure 42. Moving the transit during a stadia survey.

The distance off line when chaining 100 ft need be known only to the nearest foot (one significant figure) to determine the distance to a hundredth of a foot (five significant figures).

AZIMUTH

Ordinarily, azimuth is the horizontal angle from north to any line, measured clockwise around to 360°. Any line will have two azimuths, depending on its direction; for example, azimuth 90° is east, and 270° is west. Both are the same line, but in opposite directions. It is customary to designate directions by azimuths in stadia work, military surveys, and occasionally in land surveys. When making a stadia survey, one may assume an initial azimuth; it is equally useful whether it is truly north or not, provided later setups use the same zero.

The azimuth direction may be transferred between setups, without computations, by backsighting on the previous point (Figure 42) with telescope reversed. At the first setup, location *A*, the telescope is set with the index at zero on the graduated circle. The zero direction may be determined by a curb line, by compass, or merely by choosing a convenient direction. The lower motion is clamped while the upper motion is released during shots from this location, all of which are referred to the same zero for azimuth. When it is desired to move to *B*, a stake is driven at *B*, and an azimuth and distance read on *B*. This azimuth is locked on the instrument by tightening the upper clamp, and also leaving the lower clamp tight. You may then pick up and carry the instrument with the azimuth from *A* to *B* remaining on the plates. Elevation must be transferred also to the new location; in the notes in this manual this is done by using the transit as a level, taking a separate turning point. The location of this turning point is unimportant, so far as azimuth and distance is concerned; it need be recorded only for elevation. Some engineers use vertical angles to the new setup point to determine the new elevation; this must be done in any case if there are considerable differences in elevation (more than 5 ft up or 9 ft down).

Having recorded the turning point and the azimuth shots on point *B*, pick up the instrument and set up on point *B*. You know the azimuth from *A* to *B* is 35°, and could use a backsight of 35 + 180 = 215°, to establish your horizontal circle in the same position in relation to north as before. It is simpler to point the telescope along the 35° azimuth. You will then be pointed at 35° azimuth, and the plates will read 35°, so the zero azimuth is parallel to the azimuth at the first setup. Next, release the upper motion and proceed with azimuth readings at the new setup. You read the turning point again from this new setup to determine the new *HI*, and record this in the level notes.

This process may be continued as many times as necessary, always preserving the original azimuth. It is desirable to keep this azimuth reference for convenience in plotting. The draftsman has a drafting machine similar to a transit. It has an upper and lower motion; he sets the lower motion on your zero azimuth, and continues to plot points on the whole sheet, using the same north, just as you have done on the job.

FIELD EXERCISE NUMBER 6

The object of this exercise is to give each student an opportunity to make stadia shots, keep notes, compute elevations, and check the elevations obtained by leveling methods. By keeping the first exercise simple and confined in a small space, the instructor may more effectively supervise the group. The same rod locations may be used as were used for the level exercise (Number 4) while other students are using the same rods for the leveling exercise. In bad weather, setups may be made on a porch or covered walkway, or even near a window.

The instructor sets five rods in convenient locations, preferably so that they may be read by a level telescope. The student sets up a transit and reads one rod as a bench mark rod for elevation, recording it as shown in the notes in Figure 43, as +4.06. In these notes, the *HI* is 132.08, and the elevation from which deductions are made for measurements to the 5-ft mark on the rods, 127.1, is shown on the top of the page.

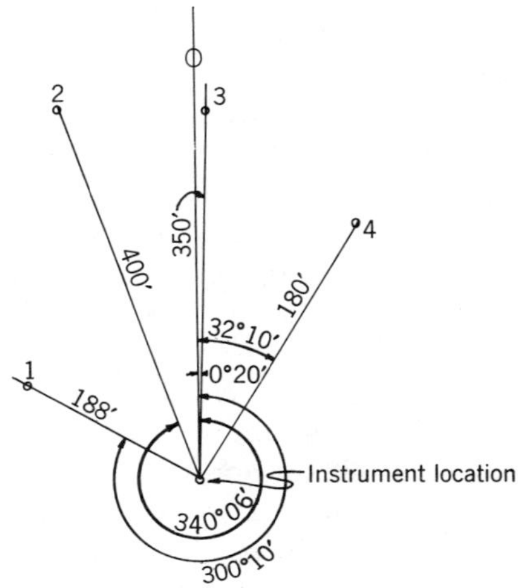

Exercise No. 6
Stadia Practice (127.1)

J. Smith π-notes 6
3-28-66

	UH	CH	LH	Azim.	V.A.	Dist.	HD/100	HD	ED/100	E.D.	Elev.
1	5.94	5.00	4.06	300-10	−02-00	188	99.88	188	−3.49	−6.6	120.5
2	7.00	5.00	3.00	340-06	+03-46	400	99.57	398	+6.56	+26.3	153.4
3	6.75	5.00	3.25	00-20	−00-26	350	99.99	350	−.76	−2.7	124.4
4	5.90	5.00	4.10	32-10	−02-30	180	99.81	180	−4.36	−7.8	119.3

Check by levels

	+	HI	−	Elev.
BM A	4.06	132.08		128.02
1			11.70	120.38
2			—	
3			7.66	124.42
4			12.71	119.37
BM A			4.05	128.03

ORIGINAL COPR. 1919 BY J. C. PARKER — MADE IN U.S.A. — TR. MK. REG. U.S. PAT. OFF.

Figure 43. Stadia practice.

The student then reads each shot as a stadia shot, recording vertical angles and crosshairs as previously explained, and computing horizontal and vertical distances. After completing the readings by vertical angles, the level shots are read on each rod, as shown under *check by levels*. The elevations shown by levels should be within 0.1 ft of those obtained by vertical angles, although many students will have greater errors than this.

This exercise will take nearly three hours for beginners, although after the first few shots, work will be fairly rapid. After these first few shots, a student should be able to proceed to making a stadia survey with little delay.

In the notes in Figure 43, shots have been numbered and a sketch drawn to show locations of shots. These notes make it possible to compare the same shots taken in two different ways. The numbers and the sketch are for practice and illustration, and are not necessary for field work, as explained in the next chapter.

The horizontal correction is insignificant for small angles. With a little practice, it is apparent when this correction may be omitted. A common error is failing to record the sign of the vertical angle, and therefore of the elevation difference; this must be recorded for vertical angle, elevation difference per 100 ft, and for total elevation difference. In these notes, a check is also shown on bench mark *A* at the end of the work; this is good practice in order to avoid the possibility of error in the first reading.

Chapter 7

STADIA SURVEY

In a small manual like this one, no attempt to show more than a small sample of typical stadia mapping situations may be made. The sample shows one type of notes; the most common variation is to keep the elevation notes on the same page as stadia shots, and to determine the height of instrument by vertical angles rather than a line of levels.

In this survey, it is assumed that an engineer has been directed to make a survey, estimate, and site plan for a retail store on a corner lot. The notes are kept to the minimum necessary for this purpose, and in this manual the material is presented as it might be accomplished in an academic course, keeping in mind that the work requiring good weather should be completed first.

FIELD NOTES—STADIA SURVEY

Figure 44 shows the notes for a simple mapping survey. The two pages are completed at the same time; page 7 shows elevations being carried forward to provide *HI* for page 8.

These notes show the party worked as follows.

1. A sketch was first made of the area to be mapped, showing only features that could not be easily described by one word in the notes. Continuous features such as the sidewalk and street were shown. Numbers 1, 2, and 3 on the sidewalk indicate the location of stadia readings which will be used to plot the sidewalk.

2. A setup is made at any selected spot from which a large number of readings can be made, and in this case, from which the bench mark can be seen. This is indicated as *A* on the sketch. It is not essential that the location of this setup be shown on the sketch, as its location would be apparent from the notes when the draftsman started to plot them. However, if the draftsman has an idea of where the setup is made, he is less

likely to start his work in a way which will run his drawing off the sheet. While set up at *A*, read the bench mark to determine the *HI* of the instrument, as if running a level line. A zero is chosen for the azimuth, by compass or by estimating, or perhaps by choosing a prominent landmark which can be located easily the following day.

3. From this first setup, stadia shots are read as in Exercise No. 6. *L* indicates that the telescope was level, as before. The first column (up and down row) on the sheet is for notes describing the point, and also to show the location of the instrument. *A* in the first line, first column, indicates that the instrument is set up at *A*. Ground shots need no labels in the first column; if no label is provided, a ground reading is assumed. Specific items to be plotted, like the oak tree in the next line, are labeled. The measurement of 10 in. indicates approximate diameter, not circumference, of the tree.

4. When all the shots which are needed or can be seen, have been taken from this location, a stake is driven at *B* and a reading taken on the ground and stake at that point. In the notes, "to *B*," indicates these readings. The distance and azimuth of this reading are used to plot the new location, but the vertical angle is used only to plot ground shots at the location of setup *B*. This vertical angle also gives an approximate check on the level reading for the *HI*; the ground under the old and new setups should have about the same difference as do the heights of instrument. After reading this stadia shot to *B*, a turning point is selected (the location is not shown in the notes) and the reading taken, 5.07, is recorded on page 7 of notes in Figure 44.

5. Leaving the reading of 100-00 firmly clamped on the plates of the transit, the transitman picks up the instrument and moves to *B*. He then flips the telescope over to the reversed position, releases the lower motion, and sights on *A*. He clamps the lower motion, flips the telescope over again to the direct position, and releases the upper motion clamp. The instrument is now oriented to the same azimuth as before. The same turning point is then read, using the telescope as a level; the reading obtained (3.83) is recorded on page 7 of the illustrated notes and the new *HI* computed. Although this *HI* is shown as entered on page 8, this is for clarity for the draftsman; the field man does not use this on page 8, as he does not compute ground shots.

After completing field work, the right half of page 8 is used for computations of distances and elevations, as shown previously, and the drawing is prepared from this data. If the transitman does not complete his work by the end of a day, he must leave marks to pick up on the next day. There should be a known point to set up on, a line of known azimuth to

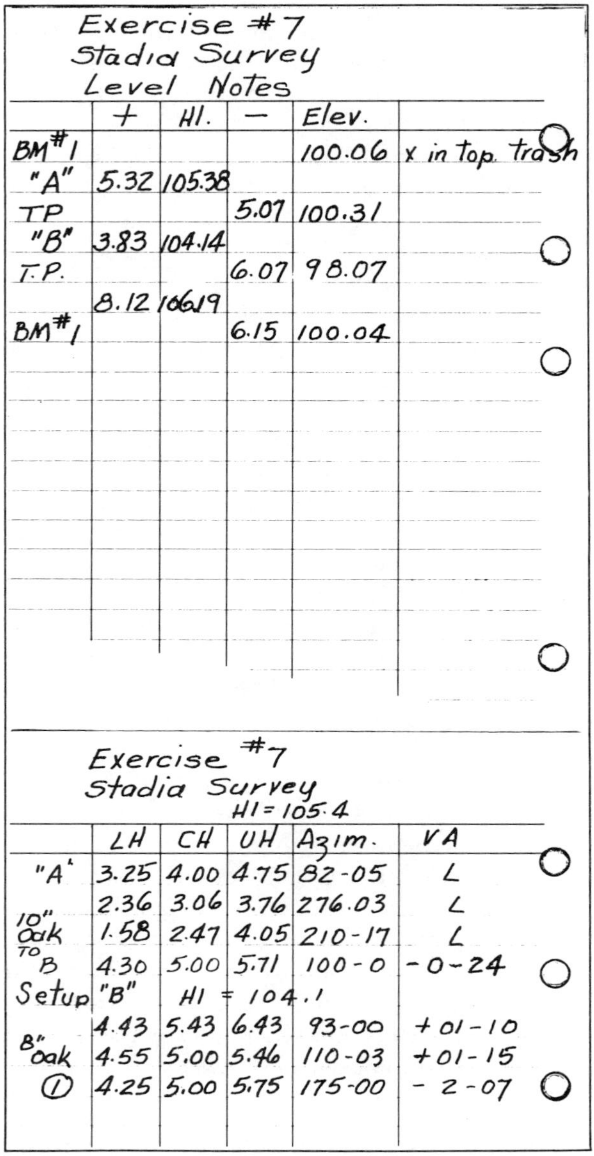

Exercise #7
Stadia Survey
Level Notes

	+	HI.	−	Elev.	
BM#1				100.06	x in top. trash
"A"	5.32	105.38			
TP			5.07	100.31	
"B"	3.83	104.14			
T.P.			6.07	98.07	
	8.12	106.19			
BM#1			6.15	100.04	

Exercise #7
Stadia Survey
HI = 105.4

	LH	CH	UH	Azim.	VA
"A"	3.25	4.00	4.75	82-05	L
	2.36	3.06	3.76	276.03	L
10" Oak	1.58	2.47	4.05	210-17	L
TO B	4.30	5.00	5.71	100-0	-0-24
Setup	"B"	HI = 104.1			
	4.43	5.43	6.43	93-00	+01-10
8" oak	4.55	5.00	5.46	110-03	+01-15
(1)	4.25	5.00	5.75	175-00	-2-07

Figure 44. Field notes—stadia survey.

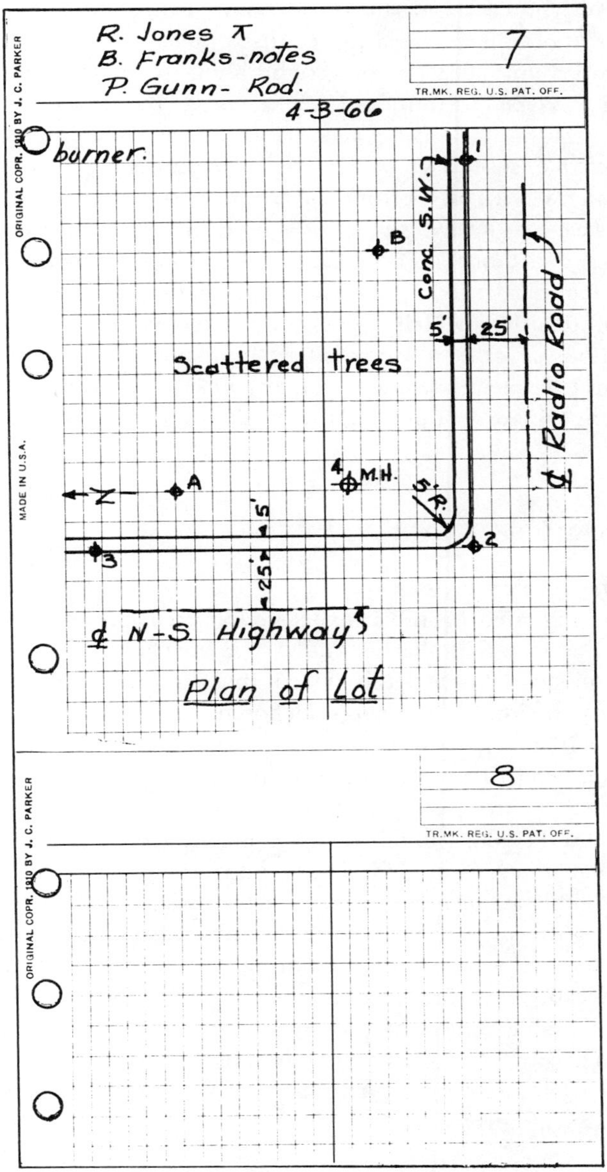

Figure 44. Continued

orient the instrument, and a turning point, bench mark or other mark from which to obtain an elevation of instrument.

The number and location of ground shots to determine topography depends on the type of ground and the accuracy required. The completed map represents the ground between shots as a straight line, so the shots, so far as possible, should be at the *breaks* or points at which the slope of ground changes, such as bottom of valleys and top of hills. When the ground surface is rounded, the shots should be taken so that the straight line from shot to shot is as often above the ground as below it; this will give the same excavation quantities as though a true picture were obtained. The detail to which topography (fixed landmarks such as trees, sidewalks, houses) is taken depends on the purpose of the map. Individual trees may not be of importance on many maps, but would be important in a location where a building is to be located so as to leave trees unharmed.

FIELD EXERCISE NUMBER 7

Each party, preferably of two or three men, is to make a stadia survey of an area designated by the instructor. As this survey is to be used for Classroom Exercises Numbers 1 and 2, the instructor should provide information on how much detail is required in the survey, and provide necessary instruction on the project chosen.

Students will take turns on notekeeper, instrument man, and rodman positions, showing their own name on the notes they record; notes need not be copied, as the party together will have a complete set of notes. Each man should take enough notes to fill a page, acting as notekeeper, if a three-man party is used. It is suggested that notes be submitted together as a single set of notes for the party, with respective pages by each man, showing the amount of work of each type completed by each student.

This exercise may be completed normally in two three-hour field periods, although by extending or reducing the area covered, it may be made to fit any available time. Computation of the notes, and plotting of the data obtained from this survey may be delayed until a further exercise, unless weather interferes, in which case the map may be completed upon completion of the survey.

CLASSROOM EXERCISE NUMBER 1

This exercise is the office work needed to complete the preparation of the topographic map for which field work was completed in Field Exercise

Number 7. Details will vary with the instructor, but the following standards are recommended:

Plot stadia survey data on tracing paper 18 × 24 in. It is economical to use a high-grade vellum rather than a cheaper thin tracing paper, because it will stand more erasing, is easier to handle, and has less shinkage. Show contours to 1-ft intervals, and plot such data as sidewalks, roads, trees, manholes, and lampposts, which must be considered in planning a new building on the site. It is suggested that the instructor post a completed drawing of the same general type on the wall, to show extent of desired detail and conventional symbols.

CLASSROOM EXERCISE NUMBER 2

Using the topographic map previously prepared in Classroom Exercise Number 1, each student is to prepare a site drawing of a project as designated by the instructor. The original contours on this drawing may be by partially tracing the topographic map, or one drawing may be used to show both old and new grades. The structure will vary according to the experience of the instructor, the type of terrain, and the emphasis of this course and other courses.

For courses with engineering emphasis, the structure may be a street intersection, a stream check, an irrigation structure, or a drainage structure. For courses with building emphasis, it may be a building site; and for architectural students, it may be part of a design problem for a new building.

A simple project may be completed in three hours, and a complicated one may take several periods.

Chapter 8

MARKING GRADES FOR CONSTRUCTION

This chapter is intended to qualify the student to set grades in the field, either on a level or sloped line, without delaying the rodman while the instrument man computes grades. The construction engineer should read the rod on the stake or other mark, make a single subtraction, and call out the stake mark to the rodman. If marks are being set or stakes driven to grade, the instrument man should know the reading to which the mark is to be set before the rodman arrives at the point. Calculations may be made for all points set from a setup, so that the rodman need not wait for calculations at each stake.

TYPES OF GRADES

The simplest grade is that of a building floor which is the same elevation throughout. However, even buildings have several floor levels, sloped floors, or floors which are curved in elevation. Streets, parking lots, yards, and pipelines are sloped, and may have a continuously changing slope. A sloped line which continuously changes grade is a *vertical curve,* usually a parabola, and seldom used except for highways. The methods for computing elevations for vertical curves are available in textbooks on route surveying, and the actual instrument work is essentially the same as for other sloped lines.

METHODS OF MARKING GRADES

Two methods of marking grade elevations are given here.

1. Mark a point that is either on the grade desired, or is a constant distance above or below the true grade.

2. First set a point, such as the top of a driven stake, and then mark on the stake the measurement up or down from this point to the true grade. These measurements are called *cut* or *fill* distances.

The grade to be set may be a slab subgrade, a finish floor, a pipe, bottom of an excavation, top of a curb, top of a manhole, highway grade, or other type of construction. It is important that the construction engineer find out directly from the foreman just what kind of grade is needed, and after the grade mark is made the engineer should again see the foreman to be sure that the foreman knows exactly what the engineer's points indicate. Failure to do this has resulted in costly errors.

MARKING A CONSTANT DISTANCE ABOVE GRADE

Marks for a floor, which are commonly made on batter boards, are usually set at actual grade when this is practical. Workmen needing other grades, such as the subgrade for a concrete slab or rough concrete surface under terrazo, make the corresponding measurements down from floor grade. Marks for fill may be made to finish grade (top of topsoil or surfacing) or to rough grade (under surfacing), and may be made directly on a long stake. Marks for a concrete wall are set at the top of the pour, and footing screeds are set to true top of concrete. Screeds for finish of slabs are set to true grade or below, far enough to compensate for screed thickness if pipe screeds are used.

MARKING CUT OR FILL ON STAKE

When ground is rough, or where the grade to be marked does not closely follow the ground line, stakes would have to be very long in order to set a mark at a constant distance above the grade. This is a common situation. For excavations, it is often necessary to set stakes for the first rough cut, although these stakes are removed or knocked out during digging; the first stakes set may be marked by cut distances on the stake, and later stakes will have marks a uniform distance above grade or at grade.

When cut or fill is marked on a stake, the point from which the measurement is made must be definite to the men working from it. Usually this point is the top of the stake, but sometimes a mark on the stake is preferred. The ground is not a dependable point of reference, since it can change by erosion caused by rain, and since the ground level at the stake can be higher or lower after a few passes of machinery excavating near it. When the ground is used as a mark for elevation, a line is drawn on the stake at ground level as assurance that there will be no change in elevation of the point used.

Cut and *fill* markings do not necessarily indicate ground changes. *Cut* may indicate the depth of earth to be removed, but not necessarily so. The machine operator is not particularly concerned with the depth of cut; this is wanted by the engineer who computes excavation. The operator wants to know the *location* of the bottom of his cut, or top of his fill. To him, *cut* is the distance from his measurement point to the bottom of the cut, and for the pipe foreman *cut* is the distance down to the working line of the pipe. *Cut* means to measure downward. The beginning of the measurement may not be the ground surface, nor the end of the measurement the bottom of the excavation. Similarly, *fill* means a measurement upward from the mark.

GRADE ROD

This explanation is for a sloped pipeline. For other types of grades, the names differ but the process is only slightly different.

To establish the *invert* elevation of a pipe (the inside bottom, usually given on pipe plans) on the job, you may do four things.

1. Compute the invert elevation from the plans.
2. Drive a stake near the line.
3. Take a shot on the stake and figure its elevation.
4. Subtract the pipe elevation from the stake elevation; this is the cut.

It is simpler and easier to figure the rod reading you should read (that is, the rod reading that would give a cut of zero) ahead of time, and then when you read the rod, subtract the rod reading from this theoretical reading, to determine the cut. The *grade rod* is this rod reading that would give a cut of zero. Since the pipe is not in place when you are setting grades for it, the grade rod is a computed reading. You can never actually read the grade rod for a pipeline underground. For a grade line above ground, you can read this grade rod, and even use it to mark the exact grade. You get this grade rod from subtracting the elevation of the invert line from the *HI*; since the *HI* changes with each setup, the grade rod must be figured again for each setup of the instrument.

STATIONING

Centerlines of long narrow projects, such as pipelines and highways, are marked with stakes every 100 ft. These are called *stations,* and are staked as described in Chapter 2. Stakes, called *pluses,* are driven between these stations where needed, and measurements to points on the line are called plus measurements. They are described by the station *plus* the distance beyond the station, as station 120 + 12.34 would be 12,012.34 ft from the point of beginning. This stationing is not important for short lines, which may be described merely by the feet and hundredths, but it would be inconvenient to use a marking of 23,456.78 ft rather than station 234 + 56.78. Stations are usually numbered from south to north, or west to east, so that the north direction may be to the top or right-hand side of the paper when the line is plotted, with numbering from left to right, as is customary.

EXAMPLE OF GRADE ROD COMPUTATIONS

The invert elevation of a sloped pipeline is read from the plans, or otherwise figured, for the first and last stake you can see for one instrument setup. The elevations at stakes between the ends may be figured by adding the difference in elevation for each interval to the preceding elevation, but it is easier to deduct the same difference from each grade rod reading. In this way, you get the grade rod readings which you are going to compare, with the actual readings on each stake.

Suppose the plans show the invert elevation at the beginning of a pipeline, on station 0 + 00, to be 100.00 ft and that the end, station 5 + 00, is 105.00 ft. The *fall,* or slope, of the line is 5 ft in 500 ft; that is 1% or 1.00 ft per 100 ft. You set up the level, and determine your initial *HI,* as in running bench levels, by taking a plus shot on a bench mark; here, you find the *HI* to be 110.00. At the zero end of the line, the grade rod will then be the difference between the invert and the *HI,* or 110.00 ft − 100.00 ft = 10.00 ft. At the station 5 + 00 end, the grade rod is 110.00 ft − 105.00 ft = 5.00 ft. We can then figure the grade rod readings between these ends by starting with 10.00 ft at the zero end, and subtracting 1.00 ft for each station.

0 + 00 grade rod	10.00
	− 1.00
1 + 00	9.00
	− 1.00
2 + 00	8.00
	− 1.00
3 + 00	7.00
	− 1.00
4 + 00	6.00
	− 1.00
5 + 00	5.00

As a check, compute the grade rod at station 5 + 00 by 110.00 − 105.00 = *5.00 check.*

REDUNDANT CHECK

Redundant means unnecessary or repetitious; a redundant check means to figure something twice to get the same answer. In the above example, you have figured the 5.00 grade rod at stations 5 + 00 by subtracting 1 ft from 10.00 five times; but you have also figured the same grade rod by subtracting the known elevation 105 from the *HI* of 110 at the end station 5 + 00, to get the same 5.00 ft for grade rod. Each of these computations checks the other; if there were any error in the adding of 1.00 at each station, there would be an error at the end, and you would immediately learn you had an error.

This redundant check is an important check in grade computations, and should be made *before* the cut or fill is marked on the stakes. To save repeating field work, you want to find an error before you have marked the stakes.

ANOTHER EXAMPLE

It is unusual for a grade to be an even percent slope. Suppose you have a grade of 2.745%, and stakes are set at closer intervals—for example, at 50 ft. However, there is no reason to attempt to mark stakes or figure grades to provide markings closer than to 0.01 ft, and more figures will increase the opportunity for error. With a 2.745% grade, a 50-ft interval will have a fall of 1.3725 ft. So if we subtract 1.37 ft each interval, and subtract an extra 0.01 ft each *fourth* interval, the result will be the same as though we retained the ten-thousandths of a foot, so far as checking at the end is concerned, and no point will be as much as 0.01 ft in error. For example,

0 + 00	20.00	2 + 00	14.51	4 + 00	9.02
	− 1.37		− 1.37		− 1.37
0 + 50	18.63	2 + 50	13.14	4 + 50	7.65
	− 1.37		− 1.37		− 1.38
1 + 00	17.26	3 + 00	11.77	5 + 00	6.28
	− 1.37		− 1.37		
1 + 50	15.89	3 + 50	10.40		
	− 1.38		− 1.37		
2 + 00	14.51	4 + 00	9.02		

Check: $2.745 \times 5 = 13.725$ $\quad 20.00 - 13.725 = 6.275$.

PIPELINE SURVEY

On engineered jobs, pipelines are originally staked by a field party employed by the designing engineer. Engineers (but usually not architects) make individual drawings of each pipeline, showing the ground elevation at each station along the line. If the job is small and the lines short, they may not show such detail. The first survey made of the line is called the *location* survey, and is made prior to construction, to determine exactly where the line is, and to determine ground elevations along the line. These ground elevations are plotted to scale, and this plot, or cross-section of the ground along the centerline of the pipe, is a *profile*. If the pipe, such as a water or gas line, need not be laid to an exact grade, location stakes may also be used for construction. Although they are lost as the trench is dug, they are no longer needed.

When it is necessary to set the pipe to exact grade, stakes must be set far enough to one side of the centerline so that they are not disturbed by excavation. The line along which these stakes are set is an *offset line,* and the stakes are *offset stakes*. For curbs and lines dug by hand, 3-ft offset, or distance from the centerline of the work to the offset stakes, is enough. If a backhoe is used, at least 5 ft offset distance is necessary, and if the trench is very deep and the material collapses during digging, the offset may be 10 ft or more. These offset stakes are set opposite all stations and other points where needed, and are used by the pipe foreman to set his own batter-boards and guide strings from which to measure to the pipe.

On a small job, such as a general contractor might be required to survey, when the location and elevation of the ends of the pipe are known before the survey is made, the location and construction survey may be made at the same time. If a record of the original ground elevation is necessary, a ground shot is taken at the centerline stake as well as on top of the grade stake to determine cut distance for the pipe. On the basis of this survey, the contractor can make a plan and profile drawing if required.

ALINEMENT NOTES

Figure 45 illustrates notes for staking a pipeline. A sketch of the area is made on the right-hand half of the page, and the left-hand side is used to indicate location of stations, type of markers, angles in the line, or other items. The sketch is not made to scale; all items to be plotted on the finished drawing are shown by dimension. On these notes, the centerline of a 12-in. pipeline is shown: when a location survey is made, a "sur-

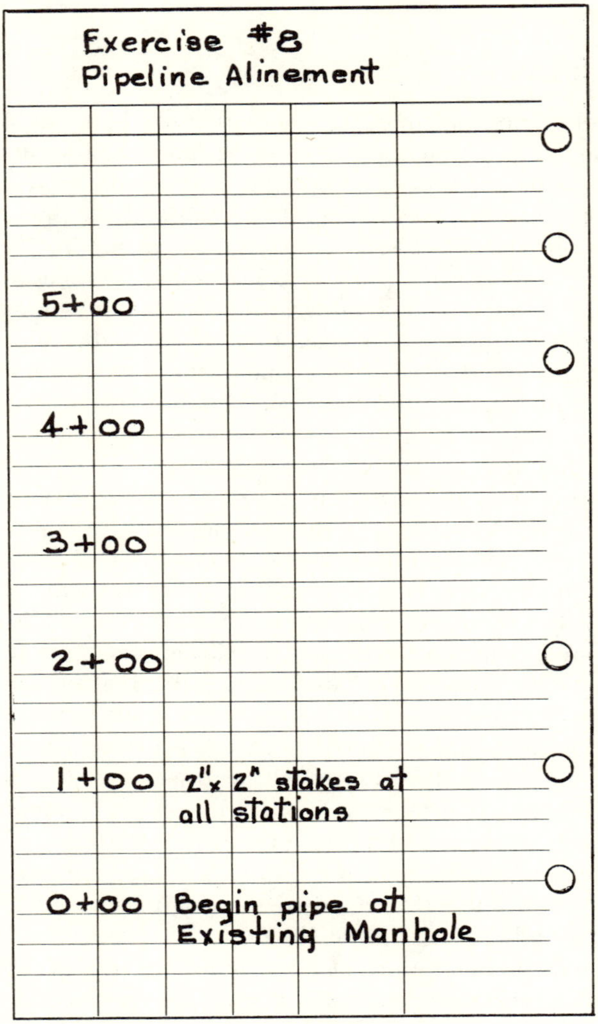

Figure 45. Field notes—pipeline alinement.

vey centerline" is used. As a rule, it is not known when the survey is made what the pipe size will be, or even the exact location of the pipe. The centerline of survey is to accumulate data, and often is not suited to be the actual centerline for construction. Notice that the beginning and end of street pavements are labeled by a plus, and the oak tree is labeled by a plus and distance. This tree is plotted on the final drawing by measuring 50 ft along the centerline and 60 ft at right angles to it.

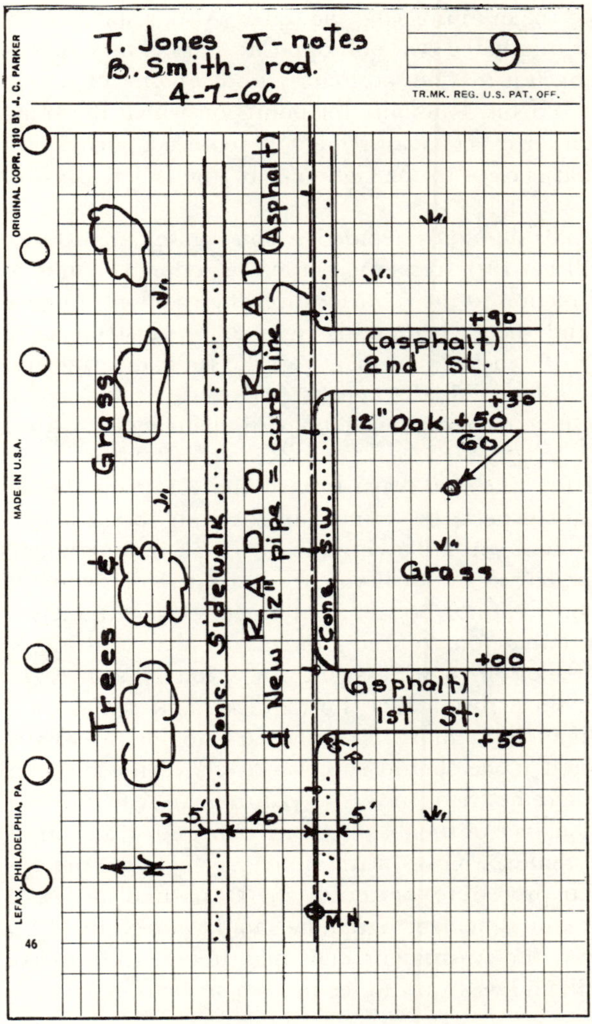

Figure 45. Continued

STAKING ALINEMENT

Alinement is staked as for chaining with stakes, except that the stakes have been marked ahead of time with the number, and they are driven with the markings toward the direction from which the measurements are being made. It is still advisable to call the station numbers, both to assure that the prenumbered stakes are used in the right order, and that there is no error in numbering the stakes. Also, a stake can be accidentally dropped, or numbered twice on opposite sides. The stakes are set by transit alinement, but it is not customary to *tackpoint,* or set exact points, on every stake. This is done only for points on which the transit is to be set up, or when an exact measurement is required, such as to find out if a building that is close to the right-of-way line is actually within the right-of-way.

Voice signals should be avoided, even at close range. The order *move to the east* will get little response while the rodman figures out which way is east; a direction *to the right* requires the rodman to figure out which is his right, and then whether his right or the instrument man's right is intended. If the instrument man uses the rodman's right or left, he must change his order method when he finds out the rodman is crosswise of the line. On a construction job, equipment noise drowns out verbal directions.

A signal for line is a wave toward the direction the rodman should move. Do not expect the rodman to see your index finger pointing; he does not know if you are giving him a signal, if your finger is still pointing from the last signal, or if you are just buttoning your shirt. A waving finger not only cannot be seen, but it shows uncertainty, as though you were not sure that you are seeing the mark. Wave with the hand and arm, with a firm motion: a wide swing for a long distance, a short hatchet swing for a short distance, a short quick hand signal for "just a hair."

The rodman must move his stake far enough to cross the line. After he has crossed it, he should successively halve the distance back to the preceding place he tried. In this way, a rodman who moves a point a foot, half of the balance of the distance each time, must be within a hundredth in no more than six moves, with an average of three moves. If he follows no particular procedure, he could take a hundred moves. When the rodman receives an alinement signal, he should move quickly and stop. If he moves slowly, the instrument man may not be able to tell that he has moved at all, and will wait to see movement before giving another signal.

DRIVING STAKES TO GRADE

The instrument man must have a signal which indicates to the rodman not only that the stake is high or low, but how much. One method of showing that the stakes must go down is to indicate on one hand the tenths, and on the other the hundredths. As the instrument man knows the conditions of driving, he may indicate by a heavy or a light stroke of one hand if the stake is to be hit hard or just tapped.

FIELD NOTES FOR PIPELINE GRADES

Figure 46 shows grade notes for a pipeline survey, which include data for two purposes.

1. The rod readings marked *G* are readings on the ground at the centerline of the pipeline. They are used during the location survey to plot the profile of the line on a drawing, and may be used to compute the quantity of excavation for payment. This plotting and computation is not done in the field, and there are therefore no computations shown for these readings in the notes.

2. The rod readings marked *S* are taken on top of the offset stakes. They are used by the pipe foreman for batterboards for pipelaying. Cut, here, does not refer to the actual earth cut, but to the distance downward from the top of the stake to the invert of the pipe.

To mark stakes as shown in these notes, there are seven steps.

1. Read a rod on the bench mark to establish the *HI,* here 110.00 ft.

2. Figure the percent slope and, therefore, the difference between adjacent stakes, if it is not given. In the example, the inverts of the pipe at each end of the pipe have been given, and the 2.744% slope is figured on the upper-right corner of the page.

3. Check the number of stakes you can see from the first setup, and figure the invert elevations for the stakes at each end of that turn (the area which can be seen from one instrument setup). This was done in the sixth column in the notes.

4. Figure the grade rod reading at one end of the turn, and proceed to figure each grade rod reading from the previous one, through the turn. Remember that if you *add* to the *elevation* from one end of the turn to the other, you must *subtract* the same amount from each *grade rod reading*. In these notes, the last stake in the turn is at station 2 + 00. To the right margin, the grade rod is checked by subtracting the invert grade for that station from the *HI* for an independent value of the grade rod.

5. Read the ground at each stake location, and record it; read the corresponding stake top, record, and immediately figure the cut distance. The cut distance is called to the rodman, who marks cut distance on each stake before he leaves it.

6. When the line extends beyond one turn, choose a turning point, move ahead, and repeat the process.

7. At completion of the work, the *HI* of the instrument must be checked by a reading on a bench mark. If this is not done, an error in any bench

Exercise #8
Pipeline Grade Survey

	+	HI	−	Elev.	Invert Gr.
BM#2	9.94	110.00		100.06	
0+0 G			13.4		
S			13.01		90.00
0+50 G			13.0		
S			12.57		2×2.744 = 5.49'
1+0 G			11.7		
S			11.06		90.00 + 5.49 = 95.49
1+50 G			11.0		
S			10.12		
2+0 G			10.4		
S			9.86		95.49
TP.			9.72	100.38	
	11.72	112.00			
2+50 G			12.0		
S			11.72		96.86
3+0 G			11.0		
S			10.03		2×2.744 = 5.49
3+50 G			11.2		
S			10.52		96.86 + 5.49 = 102.35
4+0 G			9.5		
S			8.46		
4+50 G			8.0		
S			6.32		102.35
TP			11.58	100.42	
	9.46	109.88			
BM#2			9.80	100.08	

S = top of stakes, 5' E. offset.

Figure 46. Field notes—pipeline grade survey.

mark or turning point reading will make all the work useless; checking back on the bench mark will show this error before your notes have been used by the engineer for design, or by the contractor for construction. If necessary, several turns are taken to reach a bench mark; in the notes illustrated, one turn was used. When quitting at the end of the day, a reading may be taken on any point which can be located when you start work on the following day.

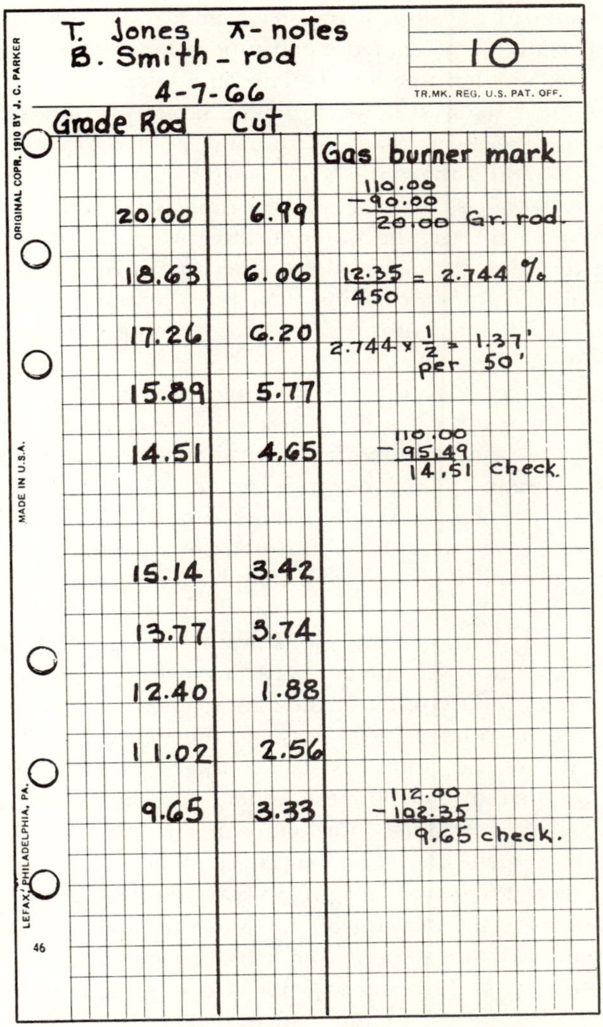

T. Jones ⊼- notes
B. Smith - rod
4-7-66

10

TR.MK. REG. U.S. PAT. OFF.

Grade Rod	Cut	Gas burner mark
		110.00 −90.00 20.00 Gr. rod
20.00	6.99	
18.63	6.06	12.35 / 450 = 2.744 %
17.26	6.20	2.744 × 1/2 = 1.37' per 50'
15.89	5.77	
14.51	4.65	110.00 −95.49 14.51 check.
15.14	3.42	
13.77	3.74	
12.40	1.88	
11.02	2.56	
9.65	3.33	112.00 −102.35 9.65 check.

ORIGINAL COPR. 1910 BY J. C. PARKER

MADE IN U.S.A.

LEFAX, PHILADELPHIA, PA.

46

Figure 46. Continued

TYPES OF PIPELINES

The preceding pipeline description would apply to a sewer. The allowable grade, and flexibility of changes, depends on the type of pipeline. The usual lines are storm sewers, sanitary sewers, water, gas and oil lines, hot water and steam lines, and electrical conduits.

Storm sewers may be laid with a slope from flat, to as high as necessary to fit the ground. If the slope is too steep, high water velocity may erode the pipe, especially if corrugated metal pipe is being used or if the water contains gravel. Smaller pipe may be laid on a steeper slope, as water will flow at the same speed as it flows through a larger pipe on a flatter slope.

For a 24 in. pipe, 0.1% grade corresponds to a velocity of about 2 ft per second (fps), and if the grade is decreased to 0.05%, the velocity is still 1.5 fps. At such grades, the slope of the pipe is not important, because only on very long lines is the water surface (which actually causes flow, rather than the pipe slope) parallel to the pipe invert. Short lengths of pipe, if full of water when carrying their rated amount, may be laid flat, as the elevation difference of incoming and outgoing water surfaces determines the flow.

Storm sewers may have dips in the line; that is, they may go down under another pipe and up again. These dips are called *inverted siphons,* although they are not siphons in the ordinary sense. At such points, material such as sand and gravel carried by the water will accumulate, so the structures must be occasionally cleaned out. This can be done with a portable pump. Inverted siphons are used for carrying irrigation water under creek beds and highways, and may be used to avoid interferences between pipelines.

Ditch grades are similar to grades of storm sewers, but the grade must be quite flat. Unless lined, few ditches may have a water velocity much over 2 ft per second without erosion of the banks; some materials, such as fine sand, will erode at 0.5 fps. Ditches must therefore be nearly flat—less than 1.0% grade. Where the ground is steeper than this, either *checks* (small dams) are installed to slow up the water, or channels are lined with asphalt or concrete. In some locations, concrete pipe has proven more economical than open ditches on very steep slopes.

Sanitary sewer grades may not greatly vary. Since high velocity causes erosion and reduces the amount of water in the pipe so that solids may not be carried away, the slope may not be too great; and if the slope is too flat, the velocity of the water will not be enough to move solid matter, thus clogging the pipe. Minimum velocity is generally 2 fps; some codes allow maximums to 15 fps. This gives slopes of about 0.5% to 1.0% minimums for different pipe sizes. The usual specification for inside buildings (plumbing codes) is ¼″ per ft, or 2.0%, but this is often reduced to 1%.

Sanitary sewers must often comply with code requirements regarding allowable slope; consequently, when there is a conflict between pipelines, the other lines can be moved without approval of public authorities, but the sanitary sewer cannot.

Inverted siphons may be installed in sanitary sewer lines, but they are expensive. Special provision must be made to maintain velocity in the low section, in order to prevent settlement of solid material.

Water, gas, and oil lines are the easiest to stake. These lines follow the contour of the ground, and the grade is determined by the minimum cover for protection against traffic and frost. Although frost penetration may be as deep as 8 ft in some areas in the United States, water mains are seldom laid this far underground. You can find out the necessary depth from your local city water department. Large mains and lines carrying water all the time rarely freeze, but small service lines must be placed the full specified distance underground. Depth of cut, therefore, may not be shown on centerline stakes, where the depth is constant.

When such lines are to be laid in areas to be later cut or filled, marking grade involves the same kind of computation as for sewers, except that the elevations are figured from the site drawings showing final grade for the ground surface. Thus the lines will be the specified distance underground after completion of site cut or fill.

Stakes for water, gas, and oil lines may be on the centerline only, with the total depth of earth cut marked on the stake, but measured from the existing surface. Since the pipe is not laid to an exact grade, the bottom of the trench is the only elevation required. These lines may make changes in grade and direction where convenient, although sewer lines must have manholes for cleanouts at turns and grade changes.

Hot water and steam lines may change direction without access manholes, but they are laid so that high or low points on the line are accessible. It is necessary to be able to bleed condensation from steam lines from the low points in the line, and the lines are laid with a slope to prevent condensate water from lying in the pipe. However, it is immaterial whether the slope is steep or flat. Expansion loops are provided at intervals, and often large concrete manholes are built to enclose these loops.

Heating lines, like sewers, are shown on profile drawings. A change in grade of a heating line may require a change in location of loops and drainage points for condensation, which are often located at the same point.

Electrical conduit may be metal, concrete, fiber, clay tile, or flexible cable. There are no restrictions in regard to slopes, drainage, or cover, so long as various minimum restrictions in regard to protection and rate of curvature are met. These seldom concern the layout engineer, and small differences in alinement are not of importance. If it is necessary to change the alinement of electrical conduit, be sure that curves are not too abrupt. For water and gas lines, this precaution is not necessary; lines may change direction with 45, or even 90° bends.

FIELD EXERCISE NUMBER 8

You are employed as engineer by a general contractor on a construction job. After the contract is signed, it is discovered that an additional drainage line is necessary during construction. The supervising engineer agrees to pay for the construction, provided that your employer furnishes all labor for the survey, and prepares the necessary drawing to authorize extra payment. Your employer, in turn, directs you to make the survey, prepare the drawing, and stake the line for the pipe foreman. Because time is short, you are obliged to stake the line for construction before the drawing is completed.

Since the alinement is already known, it is not necessary to make a location survey in advance of construction, and the procedure shown in Figure 46 may be followed. The ground shots are necessary for the later drawing, and are needed to justify the quantity of excavation for later extra payment. Since all students cannot mark the same stakes, none will mark cut marks on the stakes, but will figure them and enter them in the book as if they were marked.

This exercise may be done by a two-man party. If done twice, each student will do the complete exercise and keep notes for his own work. If there is an odd man, he can do the exercise on his own, sharing a rodman with another instrument man.

The instructor will provide data necessary to adapt this exercise to the space and time available, including the following.

1. Length of line.
2. Location of line.
3. Elevation of ends.
4. Whether entire line is to be run once or twice.
5. Extent of detail desired on plan.

For a 950-ft line, using stakes at 50-ft intervals, a two-man party should be able to run the line twice in two three-hour field periods.

The instructor will designate the bench mark to be used and the elevation assumed for it, and will provide offset stakes in advance of the level parties. In a class having a pair of experienced men, these men can stake the line in half an hour or less, especially if it is on a back line of a sidewalk where line is not required, and they will still have plenty of time to complete the exercise.

It is important that each man figure the cut amounts as he reads the stakes, and the instructor should check the first setup to make sure this is being done.

CLASSROOM EXERCISE NUMBER 3

Each student is to make a drawing of the pipeline which was surveyed as Field Exercise Number 8, similar to that shown in Figure 47. This drawing is intended to serve as guidance for the contractor's work force, and as a basis for extra payment claim to the engineer. It is not intended that the quality of the work be equal to that which would be expected of a consulting engineer who was preparing the plan for bidding and construction.

The following details are suggested.

1. Show the centerline location on plan, label beginning and end of line, and stationing.

2. Show grade and elevation of ends of line, and ground grade at beginning and end of line.

3. Drawing is on F. A. (Federal Aid) sheet 23 × 36 in., a type of drawing paper which has a profile grid below and is blank above for the plan drawing. If tracing paper is used, it is easier to draw on the back, rather than on the lined side, as the grid lines will not blot out the new lines.

4. Show sidewalks, curbs, width of streets, and cross drives in plan, and manholes and pipeline elevation in profile. Be sure the profile stations are directly below the plan stations. On the profile, plot invert line and ground, but *not* the tops of stakes. The drawing would normally be made before the stakes are marked with cut distances in the field, and the construction survey would be made after the location survey. The elevation of stakes in the field book is an adequate record; often no record of construction stakes is retained.

5. Show pipe size as 12 in., north arrow, and show scales both on plan and on profile portion of drawing, as the scales are different for the vertical portion of the profile drawing. The plan has dimensions to the same scale both ways, but the profile is usually exaggerated in elevation in order to show slopes clearly.

6. Do not forget your name. Roll the F. A. paper for submittal with the drawing on outside of roll. Use a rubber band, not a piece of drafting tape, to fasten the roll, because tape must be removed and drawings placed flat for examination; pieces of tape are hard to handle and often get lost.

7. Use scales either to suit the size of your project or as designated by the instructor.

8. Designation of pipe invert is adequate without "bottom invert," or other redundant description.

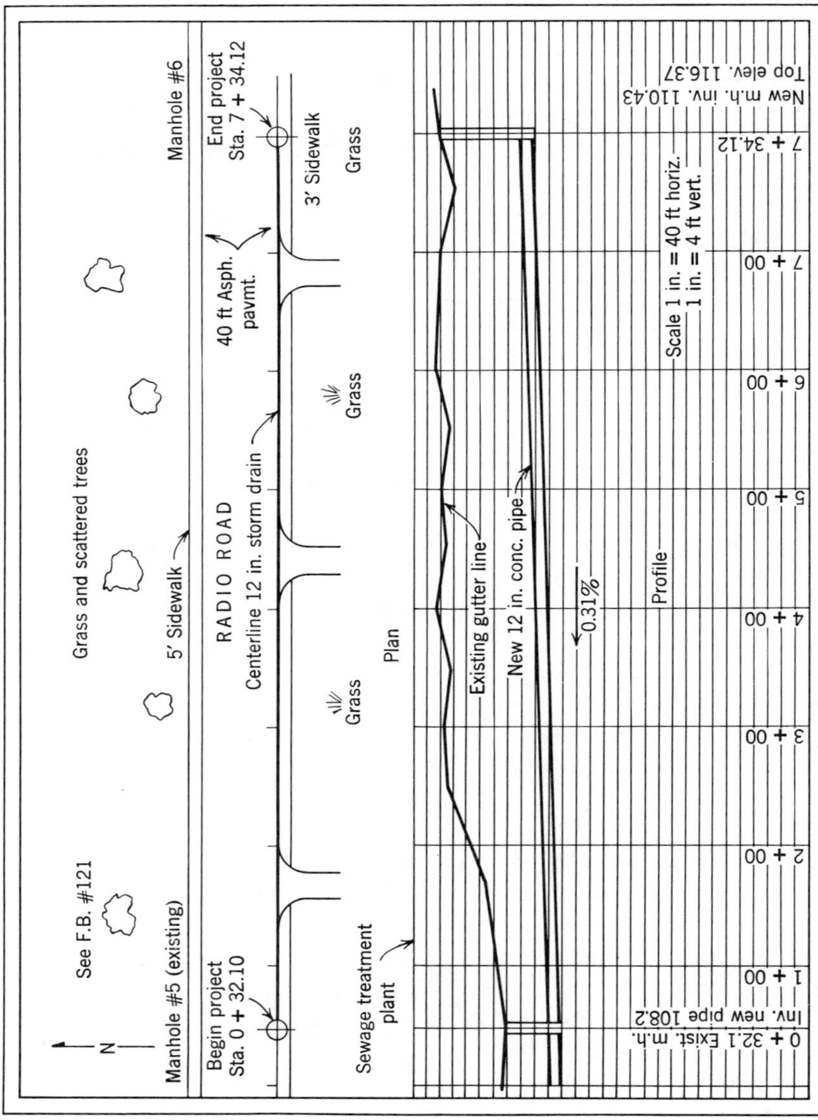

Figure 47. Plan and profile drawing.

9. The arrow should point down the slope of the pipeline to show the grade, with percent grade parallel to pipeline on profile.

10. Labels of stations should be vertical, with bottoms of printing on the station being labeled, and should be at the bottom of profile drawing.

11. Include reference to the page of your field book from which the data was obtained.

12. Elevation of ground is plotted, but it need not be labeled. If calculations of the excavation are made, they may be made with less chance of error directly from field notes. Pipeline elevations are shown at the ends only, when there are changes of grade, and at manholes.

Chapter 9

BUILDING LAYOUT

A carpenter apprentice, according to one schedule of requirements, must spend 160 hours on building layout in order to qualify as a journeyman carpenter. What then can a construction engineer do that a carpenter cannot?

In the first place, although carpenters do a quite competent job of layout of small buildings, they are rarely entrusted with larger jobs; this is done by the job superintendent himself. In the second place, this is one field where the young engineer can make himself useful; the most difficult work of the carpenter may be the simplest job for the engineer. Also, the engineer is capable of work beyond the limits of the building, being familiar with utility installations.

PROPERTY LOCATION

The contractor or builder should be furnished with a survey of the property, made by a registered land surveyor. Measurements of the lot are the smallest and simplest part of the surveyor's work.

The surveyor may be compared to a mechanic who was called to repair a complicated machine, after several others had failed. He looked over the machine carefully, picked up a hammer, and tapped it. The machine worked perfectly, and the manager was given a bill of $100.00. The manager was somewhat disturbed over the amount, and asked for a detailed accounting. The mechanic then made out a new bill.

For tapping on machine	$ 1.00
For knowing where to tap	99.00
Total	$100.00

The surveyor is this sort of skilled person. You see him measure, and assume this is what he is paid for. His skill consists in knowing *where* to measure.

CONTRACTOR'S RESPONSIBILITY

The general contractor should not survey the property on which he is to build, even if he is a qualified land surveyor; nor should he assume responsibility by engaging a land surveyor for boundary surveys. The surveyor is responsible only for using reasonable care, but the contractor may find his payment for an entire structure in danger because he has taken onto himself the responsibility of locating a structure, and has made the same error any surveyor might have made. Buildings have been built on the wrong lot, and half, or partially off, a lot. An inch error in location has resulted in a long delay of final payment.

The contractor's care starts when he determines the location of a lot. A street address is insufficient; a surveyor will not attempt to survey without a legal description, and the contractor should locate his project with the same degree of certainty. The construction contract should provide that the survey be furnished by the owner, and the contractor should be able to prove that the survey he used was in fact furnished by the owner. A single mistake of this nature can ruin a firm.

Having had a survey, it is necessary to identify the corners shown on the survey, as the points found on the ground. This may be done in the field by verifying measurements shown on the plat, or it may be necessary for the surveyor to identify his points on the job site. If you find a pipe or stake where the surveyor's plat shows it to be, do not stop hunting too soon; it is not uncommon for several different stakes to mark a property corner. It is the owner's responsibility, or the responsibility of the owner's surveyor, to determine which corner shall be used.

LENDER'S SURVEY CERTIFICATES

When a mortgage is first requested, the lender may require that a survey plat be filed. This assures him that such a lot exists, and that no one has acquired title to part of it by occupation. After the mortgage is recorded, often the same day, the lender will check the property to see if any construction work has been done or any materials have been delivered to the site. This may be combined with the boundary survey, and completed by a surveyor who reports to the lender. This check is made so that the lending agency is assured that its own mortgage takes precedence over any contractor's lien.

On large projects, or where the building is on the lot line, the lender may require that the surveyor check the actual location of the building after it has begun, to be certain there is no error in layout. Sometimes the owner will specify in this contract with the surveyor that the building corners be set at the same time as the property corners. When the building is on the property line, the contractor should have the surveyor check the building for location if the owner does not do so. Even if the contractor can measure as accurately as the surveyor, there may be minor discrepancies that would not be reported if the same engineer made all measurements.

The lender may require continued checks by a surveyor during progress of the job, as additional buildings are started, and he may even require a surveyor's certificate for each monthly payment. On smaller jobs, a single plat is made by the surveyor before the final payment is made. This plat certifies the true location of the building on the lot.

INITIAL LAYOUT

The layout engineer starts his work, as required, to furnish guides for the work being done; he has to find out what items are next needed, and must keep ahead of the various trades. Since he will have to locate nearly all points twice or more, he must plan so that he does no more work than necessary. An engineer who attempts to do a careful job may provide stakes to a much greater precision than necessary, although at the same time delaying the work. Consequently, he must be familiar with the expected progress and with the type of marks needed by each foreman.

He should obtain this information from the job superintendent, but should not expect the superintendent to give him detailed instructions. By looking for the basic information needed, rather than asking the superintendent to give him details, he will not only save the superintendent's time but also show his own ability to plan his job. The necessary stakes can be determined most readily by the job foreman, and the locations from the plans.

The layout engineer helps translate the plans to the trades. The foreman may dictate exactly how the layout is to be made, or he may merely ask for certain points, in which case the engineer must get the information for layout himself; often it is not on the plans.

Plans should agree with each other, but they must be checked to assure agreement. In the building proper, the steel anchor bolt drawings should be checked against the architect's drawings, and these against the engineering or structural drawings. Where elevations of pipework through footings are shown, these should be checked. It is unnecessary to list here the many ways in which plan differences may arise and the trouble they may cause; knowledge of such situations is one of the requirements of an experienced engineer.

The site drawings seldom offer difficulties, but the engineer should check all controlling points. Some of these are listed here.

1. Conflict of piping; piping that is shown as passing through the same location and elevation as other piping.

2. Conflict of piping and grades; piping which is shown at elevations higher than ground elevations.

3. Errors in controlling elevations, particularly sewers. All sewers on plans should be checked against existing outlets into which they are to empty.

4. Errors in critical items on the original survey, where they affect new work. Existing power poles, for example, should be checked for interference with new driveways; or if any connections, such as additions, are to be made to existing structure, the actual location of existing structure should be confirmed.

5. Seldom are ground grades poor enough that the work is seriously affected; but on one job, considerable acreage shown as land on the plans was found to be below sea level.

6. Where the engineer's employer is responsible, the true location of underground utilities in way of new construction should be confirmed. Failure to locate utilities may result in serious damage during excavation of trenches.

• LAYOUT CHECKS

Any layout dimension should be considered incomplete until there has been a redundant check. Columns should be laid out by successive measurements between centerlines, and these centerlines may be checked against the overall dimensions of the buildings to find any intermediate errors.

Lots shown as rectangular on subdivision plats are seldom actually so; sometimes the true angles are not shown on the survey plan, leading the engineer to believe that they are right angles. Therefore, it is always necessary to check the side clearances of buildings; the architect may not know the true dimensions of the lot. Surveyors are very lax (sometimes deliberately so) in furnishing property dimensions and angles. In most cases, a measurement on the drawings will show clearances of the buildings against boundaries, but a field check is often necessary.

• KINDS OF INFORMATION

Here are some of the types of layout information furnished by the engineer.

1. Elevation and location of sidewalks and curbs.
2. Layout of paved areas, streets, and parking areas.
3. Location of power poles, guy wires, and transformer banks.
4. Line and grade for sewers, electric ducts, all types of underground lines and their manholes.
5. Temporary facilities—shed locations, water, power, parking, and fence locations.
6. Line and grade for overall site grading.
7. "Hunting locations" for underground existing facilities that must be modified, removed, or avoided.
8. Wells.
9. Parking lot lighting standards.
10. Corners and elevations for buildings.

Much of the layout work may be done either by tradesmen or by the engineer, depending on the relative skills, amount of work for the engineer elsewhere, and type of job. For the general contractor, a large part of building layout usually consists of concrete work, such as setting of screed stakes, checking top of form or top of wall elevations, and marking tops of slabs inside buildings.

If properly done, foundation piers for steel columns may be poured at the correct elevation, so that no grouting is necessary. This requires careful work by the engineer, and some supervisory engineers object to the lack of an even bed for columns. Bricklayers may ask for assistance because they have considerable layout to do—both partition and wall layout and elevation layout. They set the bearing plates for beams and girders which bear on masonry, and on some jobs this layout can take up a large part of the foreman's time. It is of advantage to the young engineer to attempt as much layout for the general contract trades (assuming he is employed by the general contractor) as he can, and to this end should inquire not what the foremen want, but what he can do to help them. Even though he is not familiar with some of their problems, such as brick coursing, he can learn a great deal by encouraging foremen to use his services.

LAYOUT FOR SUBCONTRACTORS

Unless otherwise provided in contracts, subcontractors are obligated to work from whatever information is provided by previous trades. The general contractor may be obligated to furnish grading, and the utility contractor no more than he gets from the engineer—lot corners and a bench mark. In such circumstances, grading contractors may have their own survey crews, or they may hire crews from land surveyors. Interior subcontractors use existing walls and column centerlines.

It is to the advantage of the general contractor's engineer to check the subcontractor's work as much as possible. It is particularly important that he be sure the subcontractors have line and grade which are reliable, and that they all use the same lines. For example, even in a 50 ft × 50 ft bay, an electrical contractor may use one side of the bay for his grade to set floor outlets, while the plumber uses the other side to set floor plates; if each side is within 1/8 in. of the other side, and each contractor makes no more than 1/8 in. error in his location, you may have two plates adjacent to each other, but with 3/8 in. difference in elevation. No one has made a serious error, but enough have been accumulated to be significant. If the engineer sets screed stakes, he should check all other elevations of previous work in the floor, regardless of who did it.

• CONSTRUCTION LAYOUT NOTES

Conventional field notes are seldom made for building layout. Notes have no prescribed form, but the engineer should make a sketch showing the actual dimensions and angles used. This method of sketching is followed by surveyors for all situations in which there is no conventional faster method.

Notes must preserve the true angles and distances used; any form that does this is satisfactory. Many layout engineers keep a copy of the drawing on which the dimensions used are checked or circled, and any departures from the plan dimensions are shown as actual measured dimensions. These departures may be too small to be corrected, but you should show them anyway; if the distance which should be 228.11 ft is measured to be 228.12 ft, the recording of the actual measurement will be accepted as evidence of the fact that you have actually measured it. If your notes read the same as they should in all particulars, it may be assumed they are faked. Calculations used should be shown on the same sheet that is used to record the measurements; *do not use scratch sheets!*

Figure 48 shows conventional methods of labeling distances between points on a line when dimension lines are not used. Such markings are used on subdivision plats, and are adapted to dimensions on building layout.

When a print of the building drawing is used for layout notes, the engineer should carefully preserve the sheet used; often drawings are thrown away when the old ones become obsolete, or filed where they will be eventually destroyed. Some firms make a habit of requiring all old drawings be returned when new drawings are issued. This prevents obsolete drawings from being used later for construction. If the layout engineer is required to turn in his old drawings, he should see that they are preserved in a place that is accessible.

If payment is to be made from engineer's notes on the drawings, these drawings must be removed from use as soon as possible, to prevent their damage or loss. Footing elevations are often not shown on plans, but determined on the job; the actual elevations used may control the payment to the general contractor, payment to the masonry subcontractor, drawings and payments to form suppliers or reinforcing steel suppliers.

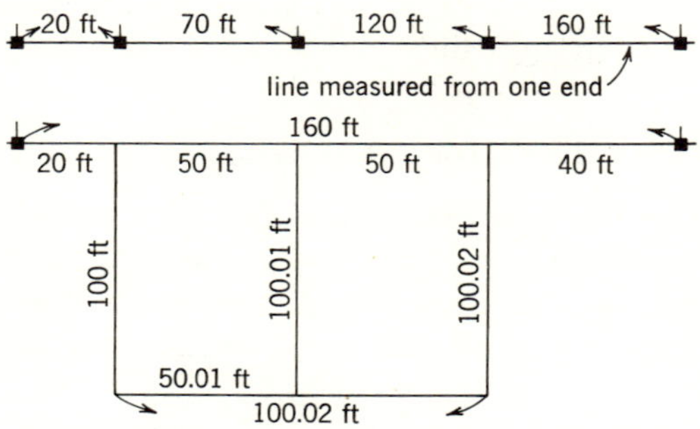

Figure 48. Designation of measurements.

REFERENCE POINTS

One of the most important jobs of a layout engineer is to preserve the location of points previously established. "Reference marks," while a general term, is usually used to denote marks or points used to reset points in plan rather than in elevation.

Points to be referenced on building construction are usually the building corners. When working to a lot line, the survey corners should be referenced if there is any question that they could possibly be disturbed. On highway work, every tenth station (1000-ft intervals) is usually referenced, as well as all curve points (beginning and end of curves, and intersections of corresponding straight lines).

The requirements for reference points are the following.

1. They must be permanent, both in material and in location. A cross in a sidewalk, or a railroad spike driven in a pavement, for example, may fulfill this requirement. When there is no such existing structure, stakes may be necessary. A nail with a bottle cap in a tree is often used, but has the disadvantage that although you can measure from such a point, you cannot set up a transit over it.

2. There must be enough reference points so that the point referenced can be reset if some of the reference points are knocked out. Figure 49 shows a method of referencing points that allows the point to be reset if any two of the four points shown remain in place.

3. Resetting points should be positive and simple. If possible, the dimensions used should be redundant; that is, it should be impossible to set an erroneous point by using one wrong dimension. If the point is

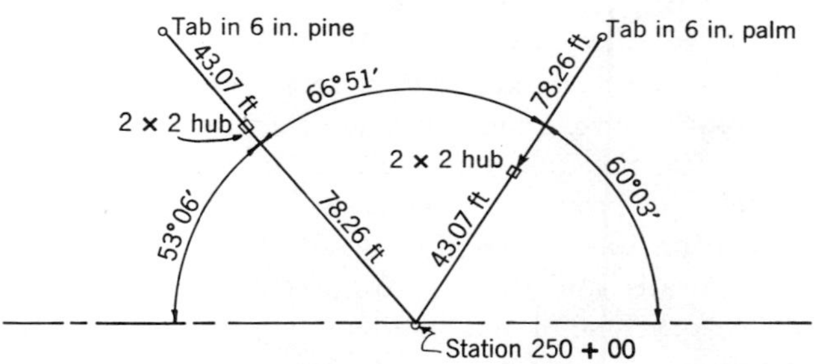

Figure 49. Reference points for highway centerline.

reset by intersection of two lines, no error in angle or distance can result in a misplaced point. Using the method of Figure 49, the point may be reset from an intersection without knowing any of the dimensions or angles. This diagram is redundant by line intersection, angle, and by dimension; all three must check.

4. If possible, no notes should be required to reset the points. That is, if all points are set at the same distance from the base point, such as 10 ft, the engineer can remember this distance. If it is not possible to use 10 ft, one may use 20 ft as the next alternative.

One such method of referencing building corners is shown on the left side of Figure 50.

The importance of reliability of reference corners depends on the importance of the point itself; lot corners, for example, are referenced much more carefully than are building corners, because the building corners can be reset from the lot corners.

Many reference points are quite temporary—perhaps used for only a

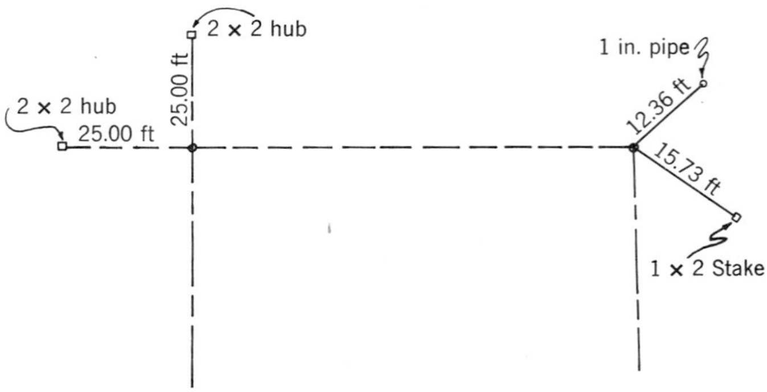

Figure 50. Typical reference points for building corners.

few hours. One may set column centerlines before pouring a concrete slab, and the points are referenced because they must be removed for the pour. The next day, the lines may be reset on the fresh concrete floor.

It is often useless to preserve points after a building is started. It is but a slight exaggeration to say that the correct dimensions are the column bolts or masonry walls, and that partition centerlines are established by the mechanical work first placed in them. A great deal of time can be wasted attempting to determine if certain work is correct when it is in place, and any possible errors are too small to require correction anyway. Insistence on correct measurements, rather than depending on work in place, is foolish, and may result in discrepancies between trades, when one trade has used the structure and another has used other points.

Figure 49 shows an elaborate set of reference points. Angles shown are unnecessary, as they are used for information and for finding the reference points, not for resetting the point. It is possible to reset a point from one of the reference points by using one of these angles, but this would be an unusual circumstance. When setting these reference points, the angle to centerline can be read in one direction from a reference line and the other reference line the other direction, making it necessary to set the zero on the transit but once. In setting these reference points, the far points are set at random, and the intermediate *hubs,* stakes driven flush with the ground and marked with a tack, are set on the same line. The angles and distances are measured, without attempting to set distances at a predetermined amount. A crew of four men should be able to reference about three such points an hour.

Figure 50 shows two simpler methods of referencing building corners. The method to the left is merely a variation of setting batter boards.

It requires that the instrument be set up over the point to be referenced, but has the advantage that no notes are necessary to reset corners; it may not be necessary to use an instrument or even a tape. The method on the right is faster; two points are set at random, and measurements recorded to the building corner. Three men are needed to reset the point, because simultaneous measurements must be made from both reference points. An important advantage of this method is that permanent points, such as sidewalks, building corners, and trees may be used. This method of choosing reference points in existing landmarks is the oldest type, and is the traditional method of referencing or *witnessing* original land corners.

BUILDER'S INSTRUMENTS

Contractors use a variety of inexpensive instruments other than those already described, such as builder's or contractor's levels, or transit-levels. A typical line is listed below, with approximate prices, and with engineering instruments included for comparison.

Table 1

Name	*1967 Price (Dollars)*	*Telescope Length (Inches)*	*Power*	*Bubble Sensitivity*
Drainage and farm level (Figure 51)	54.50	10	12	10 min./2mm.
Convertible drainage and builder's transit-level (Figure 52)	99.50	10	12	10 min./2mm.
Builder's transit level (Figure 31)	206.00	12	20	80 sec./2mm.
Builder's dumpy level (Figure 30)	159.00	12	20	80 sec./2mm.
Engineer's dumpy level (Figure 29)	380.00	18	32	20 sec./2mm.
Engineer's transit (Figure 11)	875.00	12	24	30 sec./2mm.
Self-leveling level (Figure 53)	665.00	10	32	1 sec.

The last three instruments are classed as engineer's instruments, and names are standardized. The other instruments are known by different names by different manufacturers. The drainage level may be called a farm level or utility level; the convertible level is also called a transit-level. The more reliable identification is by the bubble sensitivity for level instruments and by the vernier division for transits. The price for each class is also an identification, as the prices do not greatly vary between manufacturers of each class of instrument. Most manufacturers will furnish any instrument in the line.

Two millimeters (mm) or about 1/12 in. is the standard spacing of markings on bubble tubes, and the sensitivity indicates how far the pointing of the telescope may be changed before the bubble moves this distance. It is an indication of the reliability and accuracy of the instrument. A skilled operator can level the instrument much closer than the 2-mm interval, and should be able to consistently read the rod to within ⅓ of the indicated error for 2 mm.

The transit levels have graduated circles, and the better ones have verniers. There is but a single motion clamp, which prevents movement of the telescope. The graduated circle cannot be clamped; it is moved to a required setting by rotating it with the fingers. Since it cannot be clamped, angles cannot be repeated as with the engineer's transit.

DRAINAGE AND FARM LEVEL

The farm level can be pointed by a graduated circle but the telescope does not move up and down; consequently, it may be used only for levels, and not for setting building corners. The telescope is used for setting levels in a building at short distances; the indicated sensitivity is about 1¼ in. at 100 ft, which makes it of little use for distances over 25 to 30 ft for construction work other than grading. For establishing invert lines for ditches and piping, including plumbing, where the invert grade is 1% or more, this level is satisfactory. The error of the instrument is much less than the slope of the line, so the error is not significant. On a job requiring levels for many different kinds of work, one or two of the cheaper instruments may be left set up for the journeymen to use, leaving more expensive instruments available for other work.

CONVERTIBLE TRANSIT LEVEL

This instrument is of the same type as the farm level, as far as workmanship and accuracy is concerned, but may be used to measure horizontal angles and to give line. *Convertible* indicates that the telescope, when in use as a transit, is removed and placed in a different support on the same base. Its chief use as a transit is to line up points with one an-

Figure 51. Drainage level. (Dietzgen Co.)

Figure 52. Convertible transit level. (Dietzgen Co.)

other, for which it is sufficiently accurate in building work. It has no vernier; the expected error in pointing would be about 10′ or about the same as in level, 0.12 ft in 100 ft. If you attempt to adjust the bubble tube in these instruments to a greater accuracy than is indicated by the manufacturer of the instrument, the effort will be fruitless. The instrument will appear to be in adjustment, but then will appear to be out of adjustment when it is checked. To determine the sensitivity, move the line of sight upward until the bubble starts to move, and read the rod; then move the line of sight downward until the bubble moves again. The difference in rod reading is the greatest error which you must accept when using the instrument.

BUILDER'S TRANSIT LEVEL

This instrument is a transit with a lock which keeps the telescope from moving up and down when it is being used as a level, so that the telescope will lock in a position perpendicular to the vertical axis of the instrument. The horizontal circle has a 5′ vernier, giving a probable error of about 2′ or 0.06 ft at 100 ft. This is adequate for rough layout, or for corners which are to be checked by measurement of diagonals. The bubble accuracy will give level results accurate to within 0.01 ft at 100 ft, sufficiently accurate for ordinary building layout.

BUILDER'S DUMPY LEVEL

This is the same instrument as the transit level, but without the transit feature. Since building layout is not accurate even with the more expensive transit level, the loss in movement may not be important. For many contractors, this instrument for level, and an obsolete transit for line, are adequate for layout.

ENGINEERING INSTRUMENTS

The $875.00 price for engineer's transits refers to the domestic model. Two-thirds of engineer's transits now sold in the United States are of foreign manufacture, selling as low as $320 for a model with 20″ vernier. The self-leveling levels are presently outselling conventional bubble types in a 2 to 1 proportion, although it is complained that the pendulum which keeps the telescope level is inclined to stick, and that these instruments are affected by vibration from construction machinery.

Figure 53. Self-leveling level. (Dietzgen Co.)

LEVEL RODS

Level rods are available in a wide variety of markings, materials, folding characteristics, and removable face constructions. The removable face is preferred for training work, as rods are usually well protected and the face will wear out before the rod. In rough usage, level rods become wet both from rain and from use in making water soundings, so the wood may be ruined before the face is damaged.

For setting batter boards, a carpenter's rule may be attached to a batten for stiffness, or used by itself. The rule is the most common and immediately available tool to use as a level rod on a construction job. A folding rule with graduations to feet and hundredths is also available.

PERMANENT FORESIGHTS

A *foresight* in level notes is a minus or forward reading, but in transit work, it is a line extended forward of the instrument. For a line foresight, a flag may be set on line, a mark made, or an existing structure described. It is often desirable to be able to locate a foresight without sending a rodman to a point on the line, and marks set for such continued use are *permanent foresights.*

KINDS OF BENCH MARKS

Among items used for permanent bench marks are brass plates set in concrete, concrete monuments with brass tops, the tops of property corner pipes or monuments, crosses cut in concrete abutments, curbs or sidewalks, and nails or spikes driven horizontally into trees. Marks in trees do not change appreciably with growth of the tree, as the trunk of the tree, even when young, grows only outward. A mark placed in a young tree will be at the same elevation fifty years later.

LAYOUT MARKINGS

Ordinarily the outside wall line is used as the outside layout line of the building, and centerlines of columns are used for the interior layout lines. Many buildings, especially those with curtain walls, have exterior building lines at the outside of the columns which are different from the line at the outside of the curtain wall. Column lines frequently have small displacements in alinement, or the true straight line may be one side of columns which vary in size.

The stakes should therefore always be marked to show the layout line set, or a *guard stake* (a stake to show the location of a smaller stake that marks the point) should have the line description marked on it. "Column line" or "wall line" is not an adequate designation; if any line other than the centerline is used, it should be so stated. Centerline is designated CL, often written together as

CHECK BY JOB SUPERINTENDENT

The layout engineer should look at his work and his calculations constantly in relation to work in place to see if they look reasonable; in this way he may discover obvious errors in his computations and in the work of others.

The superintendent, in turn, can spot large errors in the work by comparing elevations and locations of various parts of the work as they are staked. Coursing of brickwork gives a good basis for checking elevation for drains and roadways; the superintendent should know the height of his eye above his feet, and by lining up with various brickwork courses can check elevations from one point to another by walking the ground and counting courses.

UNION JURISDICTION

Throughout most of the country, trade unions do not at present claim jurisdiction over layout engineers. Engineers may be involved in dispute in individual matters where unions feel that engineers are doing trade work, in cases like the following.

1. Nailing of batter boards, usually done by carpenters.
2. Layout of partitions, which is done by carpenters, or occasionally by other trades.
3. Setting of concrete screeds, which is finisher's work.
4. Stakes for sidewalk forms, which may be disputed by carpenters or finishers.
5. Checking of steel or steel alignment, which is normally done by the ironworker foreman.

In some areas, such as Southern California, operating engineers have jurisdiction of field survey work when the men are employed by the general contractor. If the surveyor is an office engineer, the claim would apply only to his field work.

In the California agreement, employees who may do construction layout are not covered if employed by the consulting engineer. That is, the union claims jurisdiction of the contractor's employees, but not necessarily of any person doing the work, which is a departure from usual union jurisdictional claims. Certain contractor's personnel are exempt from the agreement—the contractor himself, the executive, and the superintendent or office engineer who does field engineering "on occasion." Presumably, a union man would not be required unless there were a reasonable amount of full time work available for him.

Under this agreement, a party chief must be designated for a survey party of three or more members. The party chief is required to be qualified to do certain work, described in a general way in the agreement. Employees can be held responsible for loss or damage of equipment. Surveyors receive time and a half for overtime, but in assisting crafts who are receiving double time, they would get double time; this provision is a typical one for operating engineers.

In some areas, field employees of consulting engineers are required to join the professional union of which other engineering employees are members. Most surveyors are employed by consulting engineers or land surveyors, or by public agencies, and are not union members.

NOTES FOR FIELD EXERCISES 9, 10, AND 11:

It is recommended that each man keep a complete set of notes for these exercises. In this way, the instructor can check each set to see that the student understands he is to record only actual measurements, not those computed or read from a drawing.

FIELD EXERCISE NUMBER 9

Each two- or three-man party is to lay out the corners of a residence, using a transit, from the lot corners on the street, as shown in Figure 54.

Procedure is as follows.

1. Lot corners would normally be set by the land surveyor before building layout is started. If not previously set by the instructor, set these points 60 ft apart in a location designated by the instructor. Normally, the back (house side) of a sidewalk is the lot line.

2. Set up the instrument at lot corner 1, and sight on lot corner 7. You are now set on the front lot line. Set points 2 and 11; record the distances you actually measure, in accordance with the notation shown in Figure 54. If yours is a two-man crew, measure all distances from the instrument, adding the dimensions shown on the plat to determine what you will use; the front points 2 and 11 will be set 10 and 50 ft from the instrument at

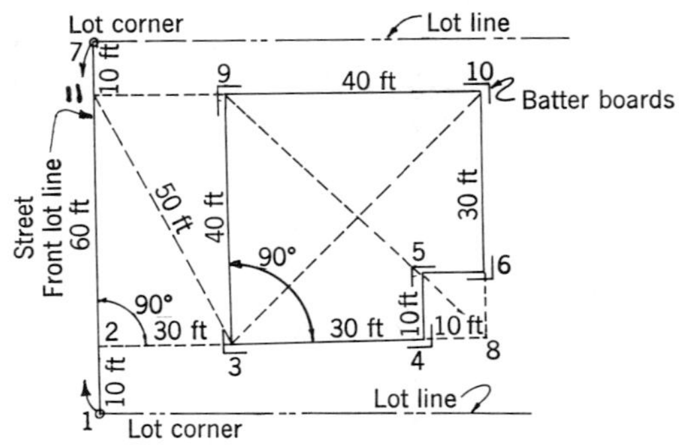

Figure 54. Building layout for Field Exercises Numbers 9, 10, 11.

the lot corner 1. Record the actual distance measured in the field book, not what you added to get the necessary measured distance. If you did not add them right, you can find the error in this way.

3. Set up on stake 2, sight on lot corner 7, and turn an angle of 90°. You are now looking down the side of the house, which is a 10-ft offset from the sideline of the lot. Since this line is established by zoning regulations and must be correct, it should be established with as little figuring as possible. If the front line were not at 90° to the side line, the distance would be slightly more than 10 ft to the first point, and the angle would not be 90°.

4. On the side line, set points 3, 4, and 8.

5. Move forward to 8, turn the 90° angle, and set points 6 and 10.

6. While still at 8, measure 10 ft from 6 and 10 ft from 4 at the same time, to set point 5. Turn a 45° angle at point 8, to see if you are set on point 5. If not, record the angle you read.

7. From point 3, set 9.

8. Set up at 10 to check angle from 8 to 9, and to see if 9 is in line with 11, as it should be. The angle at 10 should be 90°, record the actual angle you read.

9. Measure from 10 to 9, and 9 to 11, to check dimensions against what they should be. Record what you actually measure.

10. Measure diagonal 8–9 and 3–10, which should be the same. Record the actual distances you read.

Error in measurements should not be over 0.02 ft; if your error is greater than this, repeat until you find the mistake. Under job conditions, you would not check both the diagonals and the back angle at point 10.

One of these checks will be sufficient to check the layout. On flat ground, a student party should be able to do this exercise easily in two hours; if slopes require plumbing, it depends on the ability of the crew.

FIELD EXERCISE NUMBER 10

Without the transit, you are to lay out the same building as before. Stakes are alined by a tight string, and angles turned by measuring two sides of a 3:4:5 right triangle. Since it is necessary to pull the string against the stakes, which is not done in engineering work, the stakes must be carefully driven and watched for movement during the work.

Referring to Figure 54:

1. Starting from lot corners as before, stretch a string from lot corner 1 to lot corner 7 and set points 2 and 11 on line with the string.

2. Measuring 30 ft from point 2 and 50 ft from 11, set point 3. If you are part of a two-man crew, measure the lines alternately until you get the right point.

3. Pull a string over points 2 and 3 past 4 and 8, and set points 4 and 8 under the string.

4. Set point 10 by measuring from 3 and 8 as indicated, using the diagonal distance from the last exercises, or computing it. Set 6 on line between 8 and 10 by stretching the string for line.

5. Set 5 by measurement from 4 and 6.

6. Set 9 on line from 10 to 11, measuring from 11; then check by measurement from 9 to 10.

7. Check diagonals 8 to 9, and adjust stakes as necessary until all measurements check.

FIELD EXERCISE NUMBER 11

Layout building as for Field Exercise No. 9, and in addition, install batter boards at locations shown in Figure 54.

After setting corners as before do the following.

1. Drive 2 × 4 stakes, keeping back about 1 ft from the building.

2. Check corners for ground elevation, marking elevations in book. Selecting the building corner where the ground is highest, mark floor grade 12 in. above ground level.

3. Read the rod on this mark, set target at this reading (target may be a rubber band on a rod, or a mark on a batten), and mark one stake at each building corner with this grade.

4. Using the carpenter's level and a batter board, transfer elevation at the corner stake to other stakes and nail batter boards, using double-headed nails.

5. Stretch strings along each building line, putting nail in top of each batter board for strings.

Double-headed nails are used in this exercise because of the necessity for reuse of lumber. Carpenters often use saw cuts, and usually mark on the batter boards not only the building line but the footing and wall lines as well; for this exercise, the necessity for the reuse of material does not allow marking. The time required for Exercise Number 11 depends to a considerable extent on the availability of soft ground, and an area near the classroom; the quantity of material used may make moving material take more time than does the work itself.

NOTE TO INSTRUCTOR

These exercises are largely repetitive, and when it is possible to do them together some time may be saved. It has been found impractical to expect beginning students to lay out and batterboard a building, with usual academic facilities, in three hours. It should be possible to complete Exercises Number 9 and 11 together in four hours, if a continuous period is available.

Chapter 10

TRIGONOMETRY

• There are occasions when lines are either difficult to measure, or a considerable amount of measuring can be avoided by a small amount of calculation. An example is when you have measured one angle and side of a triangle and need the other two sides, and the situation makes measurement of the other sides difficult or impossible. Because calculation of the sides of a right triangle from each other is a very simple operation, there is no reason to measure a line which can be calculated.

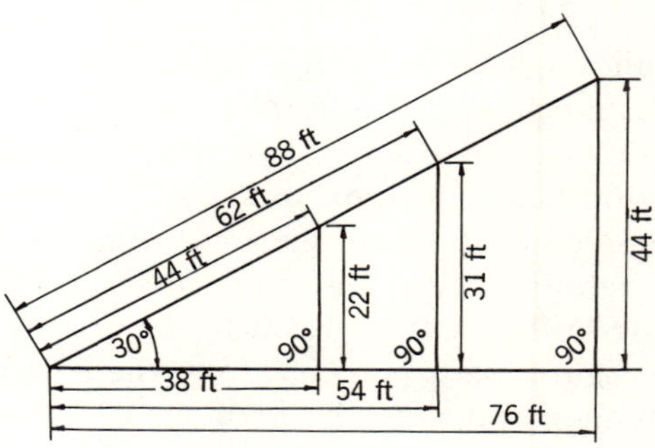

Figure 55. Right triangles.

Figure 55 shows the important feature of a right triangle, in that if one angle is the same in a number of triangles (here the angle is 30°), all the sides will increase proportionately. If the altitude increases from 22 ft to 31 ft, the sloping side, or *hypotenuse* will increase from 44 ft to 62 ft. For a 30° angle, the hypotenuse will always be twice the altitude. The base will also increase in exactly the same proportion as do the other two sides. Since the sides change in the same proportion, the ratio of one side to

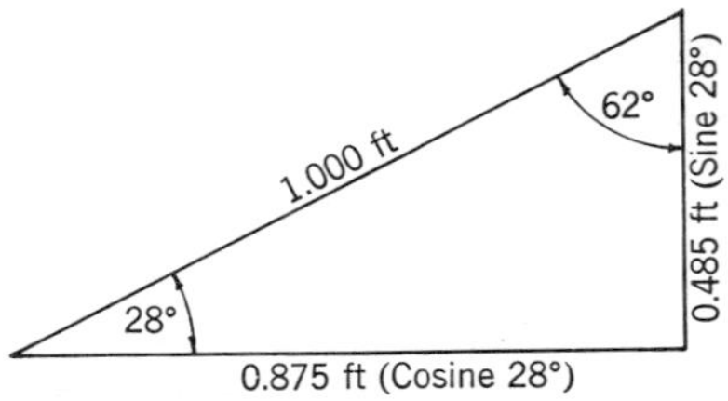

Figure 56. The sine and cosine.

another will remain the same for any angle, regardless of the dimensions of the sides. These ratios are called *functions of the angle.*

SINE AND COSINE

Figure 56 illustrates the most commonly used functions, or ratios. If you draw a right triangle with an angle of 28° and a hypotenuse of exactly 1 ft, the length of the side opposite the 28° angle will be 0.485 ft. The ratio

$$\frac{\text{opposite side}}{\text{hypotenuse}} = \frac{0.485}{1.000} = 0.485.$$

This is the *sine* of the angle of 28°. If the hypotenuse is 2 ft, the opposite side will be twice as much, or 2 ft × 0.485 ft. To find the opposite side, you therefore multiply the hypotenuse by the sine of the angle. And if you know the opposite side, and want to know the hypotenuse, you can divide the opposite side by the sine of the angle to find it.

In a similar manner, the proportion of the adjacent, or bottom, side to the hypotenuse is 0.875 ft to 1 ft, or 0.875 ft. The *cosine* of the angle is 0.875, and you can figure the hypotenuse from the adjacent side, or the adjacent side from the hypotenuse by dividing or multiplying by this ratio.

Where do you get these functions, or ratios? They are published in surveying books, mathematics manuals, and are available as loose-leaf inserts for your field book. A set of tables is illustrated in Figures 57 and 58.

If, for the 28° triangle shown in Figure 56, you had measured the bottom line, or adjacent side, and the angle, and wanted to know the opposite side, you could first divide by the cosine to get the hypotenuse, and then multiply by the sine to get the opposite side. If you can find the proportion 0.485/0.875 without having to figure it, the calculation would be quicker and simpler. Since this is a common problem, this proportion,

FOUR-PLACE TABLE OF NATURAL SINES AND COSINES

Natural
Sines and Cosines
Four-Place

TR.MK. REG. U.S. PAT. OFF.

Natural Sines from 0° to 45°

(Reading down.)

	Sine								Proportional Parts								
°	0′	10′	20′	30′	40′	50′	60′	°	1′	2′	3′	4′	5′	6′	7′	8′	9′
0	0.0000	0.0029	0.0058	0.0087	0.0116	0.0145	0.0175	89	3	6	9	12	15	17	20	23	26
1	0.0175	0.0204	0.0233	0.0262	0.0291	0.0320	0.0349	88	3	6	9	12	15	17	20	23	26
2	0.0349	0.0378	0.0407	0.0436	0.0465	0.0494	0.0523	87	3	6	9	12	15	17	20	23	26
3	0.0523	0.0552	0.0581	0.0610	0.0640	0.0669	0.0698	86	3	6	9	12	15	17	20	23	26
4	0.0698	0.0727	0.0756	0.0785	0.0814	0.0843	0.0872	85	3	6	9	12	14	17	20	23	26
5	0.0872	0.0901	0.0929	0.0958	0.0987	0.1016	0.1045	84	3	6	9	12	14	17	20	23	26
6	0.1045	0.1074	0.1103	0.1132	0.1161	0.1190	0.1219	83	3	6	9	12	14	17	20	23	26
7	0.1219	0.1248	0.1276	0.1305	0.1334	0.1363	0.1392	82	3	6	9	12	14	17	20	23	26
8	0.1392	0.1421	0.1449	0.1478	0.1507	0.1536	0.1564	81	3	6	9	12	14	17	20	23	26
9	0.1564	0.1593	0.1622	0.1650	0.1679	0.1708	0.1736	80	3	6	9	11	14	17	20	23	26
10	0.1736	0.1765	0.1794	0.1822	0.1851	0.1880	0.1908	79	3	6	9	11	14	17	20	23	26
11	0.1908	0.1937	0.1965	0.1994	0.2022	0.2051	0.2079	78	3	6	9	11	14	17	20	23	26
12	0.2079	0.2108	0.2136	0.2164	0.2193	0.2221	0.2250	77	3	6	9	11	14	17	20	23	26
13	0.2250	0.2278	0.2306	0.2334	0.2363	0.2391	0.2419	76	3	6	8	11	14	17	20	23	25
14	0.2419	0.2447	0.2476	0.2504	0.2532	0.2560	0.2588	75	3	6	8	11	14	17	20	23	25
15	0.2588	0.2616	0.2644	0.2672	0.2700	0.2728	0.2756	74	3	6	8	11	14	17	20	22	25
16	0.2756	0.2784	0.2812	0.2840	0.2868	0.2896	0.2924	73	3	6	8	11	14	17	20	22	25
17	0.2924	0.2952	0.2979	0.3007	0.3035	0.3062	0.3090	72	3	6	8	11	14	17	19	22	25
18	0.3090	0.3118	0.3145	0.3173	0.3201	0.3228	0.3256	71	3	6	8	11	14	17	19	22	25
19	0.3256	0.3283	0.3311	0.3338	0.3365	0.3393	0.3420	70	3	5	8	11	14	16	19	22	25
20	0.3420	0.3448	0.3475	0.3502	0.3529	0.3557	0.3584	69	3	5	8	11	14	16	19	22	25
21	0.3584	0.3611	0.3638	0.3665	0.3692	0.3719	0.3746	68	3	5	8	11	14	16	19	22	24
22	0.3746	0.3773	0.3800	0.3827	0.3854	0.3881	0.3907	67	3	5	8	11	13	16	19	21	24
23	0.3907	0.3934	0.3961	0.3987	0.4014	0.4041	0.4067	66	3	5	8	11	13	16	19	21	24
24	0.4067	0.4094	0.4120	0.4147	0.4173	0.4200	0.4226	65	3	5	8	11	13	16	19	21	24
25	0.4226	0.4253	0.4279	0.4305	0.4331	0.4358	0.4384	64	3	5	8	11	13	16	18	21	24
26	0.4384	0.4410	0.4436	0.4462	0.4488	0.4514	0.4540	63	3	5	8	10	13	16	18	21	23
27	0.4540	0.4566	0.4592	0.4617	0.4643	0.4669	0.4695	62	3	5	8	10	13	15	18	21	23
28	0.4695	0.4720	0.4746	0.4772	0.4797	0.4823	0.4848	61	3	5	8	10	13	15	18	20	23
29	0.4848	0.4874	0.4899	0.4924	0.4950	0.4975	0.5000	60	3	5	8	10	13	15	18	20	23
30	0.5000	0.5025	0.5050	0.5075	0.5100	0.5125	0.5150	59	3	5	8	10	13	15	18	20	23
31	0.5150	0.5175	0.5200	0.5225	0.5250	0.5275	0.5299	58	2	5	7	10	12	15	17	20	22
32	0.5299	0.5324	0.5348	0.5373	0.5398	0.5422	0.5446	57	2	5	7	10	12	15	17	20	22
33	0.5446	0.5471	0.5495	0.5519	0.5544	0.5568	0.5592	56	2	5	7	10	12	15	17	19	22
34	0.5592	0.5616	0.5640	0.5664	0.5688	0.5712	0.5736	55	2	5	7	10	12	14	17	19	22
35	0.5736	0.5760	0.5783	0.5807	0.5831	0.5854	0.5878	54	2	5	7	9	12	14	17	19	21
36	0.5878	0.5901	0.5925	0.5948	0.5972	0.5995	0.6018	53	2	5	7	9	12	14	16	19	21
37	0.6018	0.6041	0.6065	0.6088	0.6111	0.6134	0.6157	52	2	5	7	9	12	14	16	18	21
38	0.6157	0.6180	0.6202	0.6225	0.6248	0.6271	0.6293	51	2	5	7	9	11	14	16	18	20
39	0.6293	0.6316	0.6338	0.6361	0.6383	0.6406	0.6428	50	2	4	7	9	11	13	16	18	20
40	0.6428	0.6450	0.6472	0.6494	0.6517	0.6539	0.6561	49	2	4	7	9	11	13	15	18	20
41	0.6561	0.6583	0.6604	0.6626	0.6648	0.6670	0.6691	48	2	4	7	9	11	13	15	17	20
42	0.6691	0.6713	0.6734	0.6756	0.6777	0.6799	0.6820	47	2	4	6	9	11	13	15	17	19
43	0.6820	0.6841	0.6862	0.6884	0.6905	0.6926	0.6947	46	2	4	6	8	11	13	15	17	19
44	0.6947	0.6967	0.6988	0.7009	0.7030	0.7050	0.7071	45	2	4	6	8	10	12	15	17	19
°	60′	50′	40′	30′	20′	10′	0′	°	1′	2′	3′	4′	5′	6′	7′	8′	9′
	Cosine								Proportional Parts								

Natural Cosines from 45° to 90°

(Reading up.)

LEFAX,® (Marca Reg.) Philadelphia 7, Pa. Reprinted Jan., 1960

Figure 57. Natural sine and cosine (partial). (Courtesy of Lefax)

Natural
Tangent and Cotangent
Four-Place

TR.MK. REG. U.S. PAT. OFF.

FOUR-PLACE TABLE OF NATURAL TANGENTS AND COTANGENTS

Natural Tangents from 0° to 45°

(Reading down.)

	Tangent								Proportional Parts								
°	0′	10′	20′	30′	40′	50′	60′	°	1′	2′	3′	4′	5′	6′	7′	8′	9′
0	0.0000	0.0029	0.0058	0.0087	0.0116	0.0145	0.0175	89	3	6	9	12	15	17	20	23	26
1	0.0175	0.0204	0.0233	0.0262	0.0291	0.0320	0.0349	88	3	6	9	12	15	17	20	23	26
2	0.0349	0.0378	0.0407	0.0437	0.0466	0.0495	0.0524	87	3	6	9	12	15	17	20	23	26
3	0.0524	0.0553	0.0582	0.0612	0.0641	0.0670	0.0699	86	3	6	9	12	15	18	20	23	26
4	0.0699	0.0729	0.0758	0.0787	0.0816	0.0846	0.0875	85	3	6	9	12	15	18	20	23	26
5	0.0875	0.0904	0.0934	0.0963	0.0992	0.1022	0.1051	84	3	6	9	12	15	18	21	23	26
6	0.1051	0.1080	0.1110	0.1139	0.1169	0.1198	0.1228	83	3	6	9	12	15	18	21	24	27
7	0.1228	0.1257	0.1287	0.1317	0.1346	0.1376	0.1405	82	3	6	9	12	15	18	21	24	27
8	0.1405	0.1435	0.1465	0.1495	0.1524	0.1554	0.1584	81	3	6	9	12	15	18	21	24	27
9	0.1584	0.1614	0.1644	0.1673	0.1703	0.1733	0.1763	80	3	6	9	12	15	18	21	24	27
10	0.1763	0.1793	0.1823	0.1853	0.1883	0.1914	0.1944	79	3	6	9	12	15	18	21	24	27
11	0.1944	0.1974	0.2004	0.2035	0.2065	0.2095	0.2126	78	3	6	9	12	15	18	21	24	27
12	0.2126	0.2156	0.2186	0.2217	0.2247	0.2278	0.2309	77	3	6	9	12	15	18	21	24	27
13	0.2309	0.2339	0.2370	0.2401	0.2432	0.2462	0.2493	76	3	6	9	12	15	18	22	25	28
14	0.2493	0.2524	0.2555	0.2586	0.2617	0.2648	0.2679	75	3	6	9	12	16	19	22	25	28
15	0.2679	0.2711	0.2742	0.2773	0.2805	0.2836	0.2867	74	3	6	9	13	16	19	22	25	28
16	0.2867	0.2899	0.2931	0.2962	0.2994	0.3026	0.3057	73	3	6	9	13	16	19	22	25	28
17	0.3057	0.3089	0.3121	0.3153	0.3185	0.3217	0.3249	72	3	6	10	13	16	19	22	26	29
18	0.3249	0.3281	0.3314	0.3346	0.3378	0.3411	0.3443	71	3	6	10	13	16	19	23	26	29
19	0.3443	0.3476	0.3508	0.3541	0.3574	0.3607	0.3640	70	3	7	10	13	16	20	23	26	29
20	0.3640	0.3673	0.3706	0.3739	0.3772	0.3805	0.3839	69	3	7	10	13	17	20	23	27	30
21	0.3839	0.3872	0.3906	0.3939	0.3973	0.4006	0.4040	68	3	7	10	13	17	20	24	27	30
22	0.4040	0.4074	0.4108	0.4142	0.4176	0.4210	0.4245	67	3	7	10	14	17	20	24	27	31
23	0.4245	0.4279	0.4314	0.4348	0.4383	0.4417	0.4452	66	3	7	10	14	17	21	24	28	31
24	0.4452	0.4487	0.4522	0.4557	0.4592	0.4628	0.4663	65	4	7	11	14	18	21	25	28	32
25	0.4663	0.4699	0.4734	0.4770	0.4806	0.4841	0.4877	64	4	7	11	14	18	21	25	29	32
26	0.4877	0.4913	0.4950	0.4986	0.5022	0.5059	0.5095	63	4	7	11	15	18	22	25	29	33
27	0.5095	0.5132	0.5169	0.5206	0.5243	0.5280	0.5317	62	4	7	11	15	18	22	26	30	33
28	0.5317	0.5354	0.5392	0.5430	0.5467	0.5505	0.5543	61	4	8	11	15	19	23	26	30	34
29	0.5543	0.5581	0.5619	0.5658	0.5696	0.5735	0.5774	60	4	8	12	15	19	23	27	31	35
30	0.5774	0.5812	0.5851	0.5890	0.5930	0.5969	0.6009	59	4	8	12	16	20	24	27	31	35
31	0.6009	0.6048	0.6088	0.6128	0.6168	0.6208	0.6249	58	4	8	12	16	20	24	28	32	36
32	0.6249	0.6289	0.6330	0.6371	0.6412	0.6453	0.6494	57	4	8	12	16	20	25	29	33	37
33	0.6494	0.6536	0.6577	0.6619	0.6661	0.6703	0.6745	56	4	8	13	17	21	25	29	33	38
34	0.6745	0.6787	0.6830	0.6873	0.6916	0.6959	0.7002	55	4	9	13	17	21	26	30	34	39
35	0.7002	0.7046	0.7089	0.7133	0.7177	0.7221	0.7265	54	4	9	13	18	22	26	31	35	40
36	0.7265	0.7310	0.7355	0.7400	0.7445	0.7490	0.7536	53	5	9	14	18	23	27	32	36	41
37	0.7536	0.7581	0.7627	0.7673	0.7720	0.7766	0.7813	52	5	9	14	18	23	28	32	37	42
38	0.7813	0.7860	0.7907	0.7954	0.8002	0.8050	0.8098	51	5	9	14	19	24	28	33	38	43
39	0.8098	0.8146	0.8195	0.8243	0.8292	0.8342	0.8391	50	5	10	15	20	24	29	34	39	44
40	0.8391	0.8441	0.8491	0.8541	0.8591	0.8642	0.8693	49	5	10	15	20	25	30	35	40	45
41	0.8693	0.8744	0.8796	0.8847	0.8899	0.8952	0.9004	48	5	10	16	21	26	31	36	41	47
42	0.9004	0.9057	0.9110	0.9163	0.9217	0.9271	0.9325	47	5	11	16	21	27	32	37	43	48
43	0.9325	0.9380	0.9435	0.9490	0.9545	0.9601	0.9657	46	6	11	17	22	28	33	39	44	50
44	0.9657	0.9713	0.9770	0.9827	0.9884	0.9942	1.0000	45	6	11	17	23	29	34	40	46	51
°	60′	50′	40′	30′	20′	10′	0′	°	1′	2′	3′	4′	5′	6′	7′	8′	9′
	Cotangent								Proportional Parts								

Natural Cotangents from 45° to 90°

(Reading up.)

LEFAX,® (Marca Reg.) Philadelphia 7, Pa. Reprinted Nov., 1962

Figure 58. Tangents to 45°. (Courtesy of Lefax)

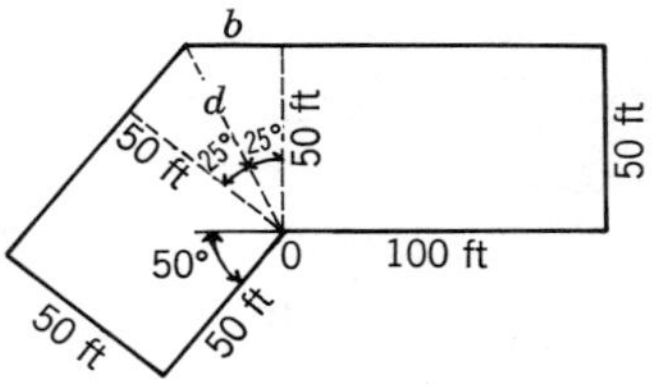

Figure 59. Oblique building.

which is the sine/cosine ratio, has also been tabulated for easy reference; this is the *tangent*. The sine and cosine are more commonly used than is the tangent. You can always divide the sine of an angle by its cosine to get the tangent, if you have no tangent tables.

ILLUSTRATIVE EXAMPLE

Architects often give front dimensions only for oblique buildings, as shown in Figure 59. This building can be staked on the job, without calculating the back line but by using the measurements given and making a trial-and-error intersection of the back lines. If you make a mistake in such a layout, there is no check on it, and this kind of layout is more work than one in which you have all the dimensions.

In this example, you are given the front and end distances 50 ft and 100 ft, and the angle the building turns, 50°. You then draw the added triangles in the corner, which represent the space added to the building at the turn. The diagonal distance, *d*, is needed so that you may set up at the corner *o* and set the back corner without setting up the back corners of the building. You need the back distance *b* in order to calculate what the length of the back line should be. If you measure this back line for a check, you do not need to set up on the back line of the building, although you may want to do so in order to check the angle. Since the turn is symmetrical, each half-turn at the corner will be 25°, or half 50°. In each of these triangles, you know the side adjacent to the 25° angle, and want to know the hypotenuse *d*. Therefore

$$d = \frac{50^\circ}{\text{cosine } 25^\circ} = \frac{50}{0.906} = 55.2 \text{ ft.}$$

Similarly, the back distance *b* is the side opposite to the 25° angle, so the distance *b* is $55.2 \times \text{sine } 25^\circ = 55.2 \times 0.423 = 23.3$ ft. The back distance can also be calculated by using the tangent of 25° and the adjacent side, since $b = 50 \times \tan 25^\circ = 50 \times 0.466 = 23.3$ ft.

These functions, given in Figures 57 and 58, are available for greater accuracy from the Lefax Company in the same page size shown, and are in the appendix of surveying textbooks. (See Recommended Readings.)

Logarithmic Function
Sines and Cosines
Four-Place

TR.MK. REG. U.S. PAT. OFF.

FOUR-PLACE LOGARITHMS OF SINES AND COSINES

Logarithms of Sines from 0° to 9°55′ (Reading down.)

°	0′	5′	10′	15′	20′	25′	30′	35′	40′	45′	50′	55′	
0	.....	7.1627	7.4637	7.6398	7.7648	7.8617	7.9408	8.0078	8.0658	8.1169	8.1627	8.2041	89
1	8.2419	8.2766	8.3088	8.3388	8.3668	8.3931	8.4179	8.4414	8.4637	8.4848	8.5050	8.5243	88
2	8.5428	8.5605	8.5776	8.5939	8.6097	8.6250	8.6397	8.6539	8.6677	8.6810	8.6940	8.7066	87
3	8.7188	8.7307	8.7423	8.7535	8.7645	8.7752	8.7857	8.7959	8.8059	8.8156	8.8251	8.8345	86
4	8.8436	8.8525	8.8613	8.8699	8.8783	8.8865	8.8946	8.9026	8.9104	8.9181	8.9256	8.9330	85
5	8.9403	8.9475	8.9545	8.9614	8.9682	8.9750	8.9816	8.9881	8.9945	9.0008	9.0070	9.0132	84
6	9.0192	9.0252	9.0311	9.0369	9.0426	9.0483	9.0539	9.0594	9.0648	9.0702	9.0755	9.0807	83
7	9.0859	9.0910	9.0961	9.1011	9.1060	9.1109	9.1157	9.1205	9.1252	9.1299	9.1345	9.1390	82
8	9.1436	9.1480	9.1525	9.1568	9.1612	9.1655	9.1697	9.1739	9.1781	9.1822	9.1863	9.1903	81
9	9.1943	9.1983	9.2022	9.2061	9.2100	9.2138	9.2176	9.2214	9.2251	9.2288	9.2324	9.2361	80
	60′	55′	50′	45′	40′	35′	30′	25′	20′	15′	10′	5′	°

Logarithms of Cosines from 80°5′ to 90° (Reading up.)

Logarithms of Sines from 10° to 45° (Reading down.)

	Sine								Porportional Parts								
°	0′	10′	20′	30′	40′	50′	60′	°	1′	2′	3′	4′	5′	6′	7′	8′	9′
10	9.2397	9.2468	9.2538	9.2606	9.2674	9.2740	9.2806	79	7	14	20	27	34	41	48	55	61
11	9.2806	9.2870	9.2934	9.2997	9.3058	9.3119	9.3179	78	6	12	19	25	31	37	43	50	56
12	9.3179	9.3238	9.3296	9.3353	9.3410	9.3466	9.3521	77	6	11	17	23	28	34	40	46	51
13	9.3521	9.3575	9.3629	9.3682	9.3734	9.3786	9.3837	76	5	11	16	21	26	32	37	42	47
14	9.3837	9.3887	9.3937	9.3986	9.4035	9.4083	9.4130	75	5	10	15	20	24	29	34	39	44
15	9.4130	9.4177	9.4223	9.4269	9.4314	9.4359	9.4403	74	5	9	14	18	23	27	32	36	41
16	9.4403	9.4447	9.4491	9.4533	9.4576	9.4618	9.4659	73	4	9	13	17	21	26	30	34	38
17	9.4659	9.4700	9.4741	9.4781	9.4821	9.4861	9.4900	72	4	8	12	16	20	24	28	32	36
18	9.4900	9.4939	9.4977	9.5015	9.5052	9.5090	9.5126	71	4	8	11	15	19	23	26	30	34
19	9.5126	9.5163	9.5199	9.5235	9.5270	9.5306	9.5341	70	4	7	11	14	18	21	25	29	32
20	9.5341	9.5375	9.5409	9.5443	9.5477	9.5510	9.5543	69	3	7	10	14	17	20	24	27	30
21	9.5543	9.5576	9.5609	9.5641	9.5673	9.5704	9.5736	68	3	6	10	13	16	19	22	26	29
22	9.5736	9.5767	9.5798	9.5828	9.5859	9.5889	9.5919	67	3	6	9	12	15	18	21	24	27
23	9.5919	9.5948	9.5978	9.6007	9.6036	9.6065	9.6093	66	3	6	9	12	15	17	20	23	26
24	9.6093	9.6121	9.6149	9.6177	9.6205	9.6232	9.6259	65	3	6	8	11	14	17	19	22	25
25	9.6259	9.6286	9.6313	9.6340	9.6366	9.6392	9.6418	64	3	5	8	11	13	16	19	21	24
26	9.6418	9.6444	9.6470	9.6495	9.6521	9.6546	9.6570	63	3	5	8	10	13	15	18	20	23
27	9.6570	9.6595	9.6620	9.6644	9.6668	9.6692	9.6716	62	2	5	7	10	12	15	17	19	22
28	9.6716	9.6740	9.6763	9.6787	9.6810	9.6833	9.6856	61	2	5	7	9	12	14	16	19	21
29	9.6856	9.6878	9.6901	9.6923	9.6946	9.6968	9.6990	60	2	4	7	9	11	13	16	18	20
30	9.6990	9.7012	9.7033	9.7055	9.7076	9.7097	9.7118	59	2	4	6	9	11	13	15	17	19
31	9.7118	9.7139	9.7160	9.7181	9.7201	9.7222	9.7242	58	2	4	6	8	10	12	14	16	19
32	9.7242	9.7262	9.7282	9.7302	9.7322	9.7342	9.7361	57	2	4	6	8	10	12	14	16	18
33	9.7361	9.7380	9.7400	9.7419	9.7438	9.7457	9.7476	56	2	4	6	8	10	11	13	15	17
34	9.7476	9.7494	9.7513	9.7531	9.7550	9.7568	9.7586	55	2	4	6	7	9	11	13	15	17
35	9.7586	9.7604	9.7622	9.7640	9.7657	9.7675	9.7692	54	2	4	5	7	9	11	12	14	16
36	9.7692	9.7710	9.7727	9.7744	9.7761	9.7778	9.7795	53	2	3	5	7	9	10	12	14	15
37	9.7795	9.7811	9.7828	9.7844	9.7861	9.7877	9.7893	52	2	3	5	7	8	10	12	13	15
38	9.7893	9.7910	9.7926	9.7941	9.7957	9.7973	9.7989	51	2	3	5	6	8	10	11	13	14
39	9.7989	9.8004	9.8020	9.8035	9.8050	9.8066	9.8081	50	2	3	5	6	8	9	11	12	14
40	9.8081	9.8096	9.8111	9.8125	9.8140	9.8155	9.8169	49	1	3	4	6	7	9	10	12	13
41	9.8169	9.8184	9.8198	9.8213	9.8227	9.8241	9.8255	48	1	3	4	6	7	9	10	11	13
42	9.8255	9.8269	9.8283	9.8297	9.8311	9.8324	9.8338	47	1	3	4	6	7	8	10	11	12
43	9.8338	9.8351	9.8365	9.8378	9.8391	9.8405	9.8418	46	1	3	4	5	7	8	9	11	12
44	9.8418	9.8431	9.8444	9.8457	9.8469	9.8482	9.8495	45	1	3	4	5	6	8	9	10	12
°	60′	50′	40′	30′	20′	10′	0′	°	1′	2′	3′	4′	5′	6′	7′	8′	9′
	Cosine								Porportional Parts								

Logarithms of Cosines from 45° to 80° (Reading up.)

LEFAX® (Marca Reg.) Philadelphia 7, Pa. Reprinted Jan., 1960

Figure 60. Logarithms of sines and cosines. (Courtesy of Lefax)

Logarithms of Cosines from 0° to 45°
(Reading down.)

	Cosine								Proportional Parts								
°	0′	10′	20′	30′	40′	50′	60′	°	1′	2′	3′	4′	5′	6′	7′	8′	9′
0	0.0000	0.0000	0.0000	0.0000	0.0000	0.0000	9.9999	89	0	0	0	0	0	0	0	0	0
1	9.9999	9.9999	9.9999	9.9999	9.9998	9.9998	9.9997	88	0	0	0	0	0	0	0	0	0
2	9.9997	9.9997	9.9996	9.9996	9.9995	9.9995	9.9994	87	0	0	0	0	0	0	0	0	0
3	9.9994	9.9993	9.9993	9.9992	9.9991	9.9990	9.9989	86	0	0	0	0	0	0	1	1	1
4	9.9989	9.9989	9.9988	9.9987	9.9986	9.9985	9.9983	85	0	0	0	0	0	1	1	1	1
5	9.9983	9.9982	9.9981	9.9980	9.9979	9.9977	9.9976	84	0	0	0	0	1	1	1	1	1
6	9.9976	9.9975	9.9973	9.9972	9.9971	9.9969	9.9968	83	0	0	0	1	1	1	1	1	1
7	9.9968	9.9966	9.9964	9.9963	9.9961	9.9959	9.9958	82	0	0	0	1	1	1	1	1	1
8	9.9958	9.9956	9.9954	9.9952	9.9950	9.9948	9.9946	81	0	0	1	1	1	1	1	2	2
9	9.9946	9.9944	9.9942	9.9940	9.9938	9.9936	9.9934	80	0	0	1	1	1	1	1	2	2
10	9.9934	9.9931	9.9929	9.9927	9.9924	9.9922	9.9919	79	0	0	1	1	1	1	2	2	2
11	9.9919	9.9917	9.9914	9.9912	9.9909	9.9907	9.9904	78	0	1	1	1	1	2	2	2	2
12	9.9904	9.9901	9.9899	9.9896	9.9893	9.9890	9.9887	77	0	1	1	1	1	2	2	2	3
13	9.9887	9.9884	9.9881	9.9878	9.9875	9.9872	9.9869	76	0	1	1	1	2	2	2	2	3
14	9.9869	9.9866	9.9863	9.9859	9.9856	9.9853	9.9849	75	0	1	1	1	2	2	2	3	3
15	9.9849	9.9846	9.9843	9.9839	9.9836	9.9832	9.9828	74	0	1	1	1	2	2	2	3	3
16	9.9828	9.9825	9.9821	9.9817	9.9814	9.9810	9.9806	73	0	1	1	1	2	2	3	3	3
17	9.9806	9.9802	9.9798	9.9794	9.9790	9.9786	9.9782	72	0	1	1	2	2	2	3	3	4
18	9.9782	9.9778	9.9774	9.9770	9.9765	9.9761	9.9757	71	0	1	1	2	2	3	3	3	4
19	9.9757	9.9752	9.9748	9.9743	9.9739	9.9734	9.9730	70	0	1	1	2	2	3	3	4	4
20	9.9730	9.9725	9.9721	9.9716	9.9711	9.9706	9.9702	69	0	1	1	2	2	3	3	4	4
21	9.9702	9.9697	9.9692	9.9687	9.9682	9.9677	9.9672	68	0	1	1	2	2	3	3	4	4
22	9.9672	9.9667	9.9661	9.9656	9.9651	9.9646	9.9640	67	1	1	2	2	3	3	4	4	5
23	9.9640	9.9635	9.9629	9.9624	9.9618	9.9613	9.9607	66	1	1	2	2	3	3	4	4	5
24	9.9607	9.9602	9.9596	9.9590	9.9584	9.9579	9.9573	65	1	1	2	2	3	3	4	5	5
25	9.9573	9.9567	9.9561	9.9555	9.9549	9.9543	9.9537	64	1	1	2	2	3	4	4	5	5
26	9.9537	9.9530	9.9524	9.9518	9.9512	9.9505	9.9499	63	1	1	2	3	3	4	4	5	6
27	9.9499	9.9492	9.9486	9.9479	9.9473	9.9466	9.9459	62	1	1	2	3	3	4	5	5	6
28	9.9459	9.9453	9.9446	9.9439	9.9432	9.9425	9.9418	61	1	1	2	3	3	4	5	5	6
29	9.9418	9.9411	9.9404	9.9397	9.9390	9.9383	9.9375	60	1	1	2	3	4	4	5	6	6
30	9.9375	9.9368	9.9361	9.9353	9.9346	9.9338	9.9331	59	1	1	2	3	4	4	5	6	7
31	9.9331	9.9323	9.9315	9.9308	9.9300	9.9292	9.9284	58	1	2	2	3	4	5	5	6	7
32	9.9284	9.9276	9.9268	9.9260	9.9252	9.9244	9.9236	57	1	2	2	3	4	5	6	6	7
33	9.9236	9.9228	9.9219	9.9211	9.9203	9.9194	9.9186	56	1	2	2	3	4	5	6	7	7
34	9.9186	9.9177	9.9169	9.9160	9.9151	9.9142	9.9134	55	1	2	3	3	4	5	6	7	8
35	9.9134	9.9125	9.9116	9.9107	9.9098	9.9089	9.9080	54	1	2	3	4	5	5	6	7	8
36	9.9080	9.9070	9.9061	9.9052	9.9042	9.9033	9.9023	53	1	2	3	4	5	6	7	8	8
37	9.9023	9.9014	9.9004	9.8995	9.8985	9.8975	9.8965	52	1	2	3	4	5	6	7	8	9
38	9.8965	9.8955	9.8945	9.8935	9.8925	9.8915	9.8905	51	1	2	3	4	5	6	7	8	9
39	9.8905	9.8895	9.8884	9.8874	9.8864	9.8853	9.8843	50	1	2	3	4	5	6	7	8	9
40	9.8843	9.8832	9.8821	9.8810	9.8800	9.8789	9.8778	49	1	2	3	4	5	6	8	9	10
41	9.8778	9.8767	9.8756	9.8745	9.8733	9.8722	9.8711	48	1	2	3	4	6	7	8	9	10
42	9.8711	9.8699	9.8688	9.8676	9.8665	9.8653	9.8641	47	1	2	3	5	6	7	8	9	10
43	9.8641	9.8629	9.8618	9.8606	9.8594	9.8582	9.8569	46	1	2	4	5	6	7	8	10	11
44	9.8569	9.8557	9.8545	9.8532	9.8520	9.8507	9.8495	45	1	2	4	5	6	7	9	10	11
°	60′	50′	40′	30′	20′	10′	0′	°	1′	2′	3′	4′	5′	6′	7′	8′	9′
	Sine								Proportional Parts								

Logarithms of Sines from 45° to 90°
(Reading up.)

Figure 60. Continued

NATURAL AND LOGARITHMIC FUNCTIONS

The functions shown above (sine, cosine, and tangent) are arithmetical ratios, or *natural* functions. They are used in this form for short distances, occasionally when the arithmetic is not too laborious, and in offices where computers are available. When lines must be calculated to the same degree of precision as they can be measured to in the field, or about a half foot in a mile, the tables used must be to six or seven significant figures. Frequent multiplications and subtractions must be made for numbers like 10,356.27 × 0.4457897. To make these arithmetical operations easier, the *logarithmic* system of computation is used.

LOGARITHMS

A logarithm is a kind of code number. Suppose you have the preceding multiplication 10,356.27 × 0.4557897. Suppose you can find a code number for each, add the code numbers, and then look in your code book to convert the total of the code numbers back to a product. You will have had to convert three numbers to these codes, but will have an addition instead of subtraction. This is *logarithmic computation,* and can be used for any multiplication. If you want to divide instead of multiply, the code numbers, or *logarithms* are subtracted instead of added.

But looking up these logarithms for each number is in many cases more work than multiplying the numbers in the first place. However, we have to look up the functions anyway; if we need sine 25°, why have one table for the natural function, and then have to go to another to find the logarithm of this function? This can be combined in one table, so you can find the logarithm of the function instead of the function itself. This is a table of *logarithmic functions,* Figures 60 and 61. A more accurate table of functions appears in the Appendix of this manual.

LOGARITHMS OF NUMBERS

What are these code numbers that are called logarithms? They are the exponent of 10 which is necessary in order to get the number. Since 10^1 indicates merely 10, then 1 is the logarithm of 10.

To find out the logarithm for other numbers, note that when a number is multiplied by itself, a 2 exponent expresses the multiplication: $10 \times 10 = 10^2$ or $10 \times 10 \times 10 \times 10 = 10^4$, which is the conventional way of showing that a number has been multiplied by itself..

To divide one number by another

$$\frac{10 \times 10 \times 10}{10 \times 10} = \frac{10^3}{10^2} = 10^{(3-2)} = 10$$

Logarithmic Function
Tangents & Cotangents
Four-Place

TR.MK. REG. U.S. PAT. OFF.

FOUR-PLACE LOGARITHMS OF TANGENTS AND COTANGENTS

Logarithms of Tangents from 0° to 9°55′ (Reading down.)

°	0′	5′	10′	15′	20′	25′	30′	35′	40′	45′	50′	55′	
0	......	7.1627	7.4637	7.6398	7.7648	7.8617	7.9409	8.0078	8.0658	8.1170	8.1627	8.2041	89
1	8.2419	8.2767	8.3089	8.3389	8.3669	8.3932	8.4181	8.4416	8.4639	8.4851	8.5053	8.5246	88
2	8.5431	8.5608	8.5779	8.5943	8.6101	8.6254	8.6401	8.6544	8.6682	8.6815	8.6945	8.7071	87
3	8.7194	8.7313	8.7429	8.7542	8.7653	8.7760	8.7865	8.7967	8.8067	8.8165	8.8261	8.8355	86
4	8.8446	8.8536	8.8624	8.8711	8.8795	8.8878	8.8960	8.9040	8.9118	8.9196	8.9272	8.9346	85
5	8.9420	8.9492	8.9563	8.9633	8.9701	8.9769	8.9836	8.9901	8.9966	9.0030	9.0093	9.0155	84
6	9.0216	9.0277	9.0336	9.0395	9.0453	9.0510	9.0567	9.0622	9.0678	9.0732	9.0785	9.0839	83
7	9.0891	9.0943	9.0995	9.1045	9.1096	9.1145	9.1194	9.1243	9.1291	9.1338	9.1385	9.1432	82
8	9.1478	9.1524	9.1569	9.1613	9.1658	9.1702	9.1745	9.1788	9.1831	9.1873	9.1915	9.1956	81
9	9.1997	9.2038	9.2078	9.2118	9.2158	9.2197	9.2236	9.2275	9.2313	9.2351	9.2389	9.2426	80
	60′	55′	50′	45′	40′	35′	30′	25′	20′	15′	10′	5′	°

Logarithms of Cotangents from 80°5′ to 90° (Reading up.)

Logarithms of Tangents from 10° to 45° (Reading down.)

	Tangent								Proportional Parts								
°	0′	10′	20′	30′	40′	50′	60′	°	1′	2′	3′	4′	5′	6′	7′	8′	9′
10	9.2463	9.2536	9.2609	9.2680	9.2750	9.2819	9.2887	79	7	14	21	28	35	42	49	57	64
11	9.2887	9.2953	9.3020	9.3085	9.3149	9.3212	9.3275	78	6	13	19	26	32	39	45	52	58
12	9.3275	9.3336	9.3397	9.3458	9.3517	9.3576	9.3634	77	6	12	18	24	30	36	42	48	54
13	9.3634	9.3691	9.3748	9.3804	9.3859	9.3914	9.3968	76	6	11	17	22	28	33	39	45	50
14	9.3968	9.4021	9.4074	9.4127	9.4178	9.4230	9.4281	75	5	10	16	21	26	31	37	42	47
15	9.4281	9.4331	9.4381	9.4430	9.4479	9.4527	9.4575	74	5	10	15	20	25	29	34	39	44
16	9.4575	9.4622	9.4669	9.4716	9.4762	9.4808	9.4853	73	5	9	14	19	23	28	32	37	42
17	9.4853	9.4898	9.4943	9.4987	9.5031	9.5075	9.5118	72	4	9	13	18	22	26	31	35	40
18	9.5118	9.5161	9.5203	9.5245	9.5287	9.5329	9.5370	71	4	8	13	17	21	25	29	34	38
19	9.5370	9.5411	9.5451	9.5491	9.5531	9.5571	9.5611	70	4	8	12	16	20	24	28	32	36
20	9.5611	9.5650	9.5689	9.5727	9.5766	9.5804	9.5842	69	4	8	12	15	19	23	27	31	35
21	9.5842	9.5879	9.5917	9.5954	9.5991	9.6028	9.6064	68	4	7	11	15	19	22	26	30	33
22	9.6064	9.6100	9.6136	9.6172	9.6208	9.6243	9.6279	67	4	7	11	14	18	21	25	29	32
23	9.6279	9.6314	9.6348	9.6383	9.6417	9.6452	9.6486	66	3	7	10	14	17	21	24	28	31
24	9.6486	9.6520	9.6553	9.6587	9.6620	9.6654	9.6687	65	3	7	10	13	17	20	23	27	30
25	9.6687	9.6720	9.6752	9.6785	9.6817	9.6850	9.6882	64	3	6	10	13	16	20	23	26	29
26	9.6882	9.6914	9.6946	9.6977	9.7009	9.7040	9.7072	63	3	6	9	13	16	19	22	25	28
27	9.7072	9.7103	9.7134	9.7165	9.7196	9.7226	9.7257	62	3	6	9	12	15	19	22	25	28
28	9.7257	9.7287	9.7317	9.7348	9.7378	9.7408	9.7438	61	3	6	9	12	15	18	21	24	27
29	9.7438	9.7467	9.7497	9.7526	9.7556	9.7585	9.7614	60	3	6	9	12	15	18	21	24	26
30	9.7614	9.7644	9.7673	9.7701	9.7730	9.7759	9.7788	59	3	6	9	12	14	17	20	23	26
31	9.7788	9.7816	9.7845	9.7873	9.7902	9.7930	9.7958	58	3	6	9	11	14	17	20	23	26
32	9.7958	9.7986	9.8014	9.8042	9.8070	9.8097	9.8125	57	3	6	8	11	14	17	20	22	25
33	9.8125	9.8153	9.8180	9.8208	9.8235	9.8263	9.8290	56	3	5	8	11	14	16	19	22	25
34	9.8290	9.8317	9.8344	9.8371	9.8398	9.8425	9.8452	55	3	5	8	11	14	16	19	22	24
35	9.8452	9.8479	9.8506	9.8533	9.8559	9.8586	9.8613	54	3	5	8	11	13	16	19	21	24
36	9.8613	9.8639	9.8666	9.8692	9.8718	9.8745	9.8771	53	3	5	8	11	13	16	18	21	24
37	9.8771	9.8797	9.8824	9.8850	9.8876	9.8902	9.8928	52	3	5	8	10	13	16	18	21	24
38	9.8928	9.8954	9.8980	9.9006	9.9032	9.9058	9.9084	51	3	5	8	10	13	16	18	21	23
39	9.9084	9.9110	9.9135	9.9161	9.9187	9.9212	9.9238	50	3	5	8	10	13	15	18	21	23
40	9.9238	9.9264	9.9289	9.9315	9.9341	9.9366	9.9392	49	3	5	8	10	13	15	18	20	23
41	9.9392	9.9417	9.9443	9.9468	9.9494	9.9519	9.9544	48	3	5	8	10	13	15	18	20	23
42	9.9544	9.9570	9.9595	9.9621	9.9646	9.9671	9.9697	47	3	5	8	10	13	15	18	20	23
43	9.9697	9.9722	9.9747	9.9772	9.9798	9.9823	9.9848	46	3	5	8	10	13	15	18	20	23
44	9.9848	9.9874	9.9899	9.9924	9.9949	9.9975	0.0000	45	3	5	8	10	13	15	18	20	23
°	60′	50′	40′	30′	20′	10′	0′	°	1′	2′	3′	4′	5′	6′	7′	8′	9′
	Cotangent								Proportional Parts								

Logarithms of Cotangents from 45° to 80° (Reading up.)

LEFAX® (Marca Reg.) Philadelphia 7, Pa. Reprinted Jan., 1960

Figure 61. Logarithms of tangents 0–45°. (Courtesy of Lefax)

This division is by cancellation of 10's below the line and those above it, and the 2 and 3 logarithms. The subtraction (3 − 2) is a method of showing how many 10's have been cancelled. The suubtraction of 2 from 3 is a *logarithmic division,* or subtraction of logarithms, and is written:

$$\begin{aligned} \log 1000 &= 3.0000 \\ \log\ \ 100 &= \underline{2.0000} \\ \log\ \ \ 10 &= 1.0000 \end{aligned}$$

Since you get the 1.0000 and from this figure out what the number is, the 10 is called the *antilogarithm* (abbreviated *antilog*). It means that the number of which the other is a logarithm, as above would be 1.0000 = antilog 10.

This is a simple example, using logarithms which are whole numbers. The logarithm of any number consisting of 1 followed by zeros is one less than the number of digits in the number, as log 1 = 0, log 10 = 1, log 100 = 2. For logarithms and antilogarithms other than these, tables are used to determine the exact logarithm. These tables are shown in Figure 62 and in the Appendix.

The logarithm of 100 is exactly 2 and of 1000 is exactly 3. Any number between 100 and 1000 will have a logarithm between 2 and 3, or 2 + something. This 2 is the whole number of the logarithm or the *characteristic.* This is not in the tables, but you calculate it by subtracting 1 from the number of digits before the decimal point; any number between 100 and 1000, such as 175, 357, 876, will have three digits and the characteristic of the logarithm will be 2. If the number is 23, the characteristic is 1; if it is 2, the characteristic is 0. Similarly, the characteristic of 0.1 or 0.12 is −1. You should remember that you must move the decimal point one digit to the right to get the whole number, and that the characteristic is the number of digits you must move the decimal point. The characteristic of 0.033 is −2, since the decimal point must be moved two digits to the right to get a whole number.

Remember that in determining the characteristic, the digits to the right of the decimal point do not count. The characteristic of 325, 325.1, 325.13, or 325.136 are all 2.

In order to save space, the characteristic is not shown in logarithmic tables of numbers. The value in the tables is the part of the logarithm to the right of the decimal point, or the *mantissa.* If you look in the table for the logarithm of 10, you find 0000; this indicates that the logarithm characteristic (which is not shown and must be figured), or 2, has the balance of the figures 0.0000, so that the logarithm is 2.0000. If you look for

10 or 100, you get the same mantissa. Consequently, there is no decimal point shown in the logarithmic tables of numbers; the mantissa of 4.7, 47, 470, is the same. Only the characteristic differs. This makes it possible to put all logarithms of all numbers on a few sheets. For instance

$$\log 5.00 = 0.6990$$
$$\log 50.0 = 1.6990$$
$$\log 500 \ = 2.6990$$

PROPORTIONAL PARTS

The table shows the logarithm of a number of two digits, such as 37, directly, in Figure 62 and provides a column to read the logarithm of a 3-digit number. For example, to read the mantissa for 372, go to the line for 37 and then to the right for the mantissa under the column headed "2." The mantissa is *5705* and hence the logarithm of 373 is *2.5705.*

The principle of proportional parts may be illustrated by a familiar example. If you have weight tables which show the weight of a 5-ft man as 150 lb, and a 6-ft man as 162 lb, what would be the weight of a man 5 ft 6 in tall? Since the difference between the 5-ft and 6-ft man is 12 lb, you have 6 in. more than the 150 lb man, or $6/12 \times 12$ lb $= 6$ lb. The 5 ft 6 in. man weighs 6 lb more than a 5-ft man, so you add 6 lb to the 150 lb of a 5-ft man, to get 156 lb. The 6/12 is the *proportional part* of the difference between the two. You know that weight does not vary directly in proportion to height; the taller man weighs more per foot of height than does the small man. Consequently, this sort of calculation is not exact in this case, nor is it exact for logarithms. It is close enough for our purposes, however.

Now suppose that you want the logarithm of 372.3. You have the logarithm of 372 on the chart, and also the logarithm of 373. You need a value which is 3/10 of the distance from the logarithm of 372 (5705) to the logarithm of 373 (5717). The difference is 12, so 3/10 of 12 is 3.6. Since we are using but one digit, call it 4. 5705 and 4 are 5709, the mantissa of 372.3. This 5709, like all values on the table, is really 0.5709, and the characteristic is 2, so the logarithm of 372.3 is 2.5709.

But if you use the "proportional parts" columns on the right of the table, you do not need to subtract the mantissas and multiply by the proportion. Since you want 3 greater than 372, look in the proportional parts column under "3" and you read "3" as the amount to be added. Adding this to 5705, you then get 5708 instead of 5709 as above. This is near enough, and much easier. It is not exact, since these proportional parts are of values from 370 to 379, or differences from 11 to 12. You do

<table>
<tr><td>Logarithms</td></tr>
<tr><td>Common, 4-place</td></tr>
<tr><td>Proportional Parts</td></tr>
</table>

TR. MK. REG. U.S. PAT. OFF.

LOGARITHMS OF NUMBERS
PROPORTIONAL PARTS

	0	1	2	3	4	5	6	7	8	9	Proportional Parts								
											1	2	3	4	5	6	7	8	9
10	0000	0043	0086	0128	0170	0212	0253	0294	0334	0374	4	8	12	17	21	25	29	33	37
11	0414	0453	0492	0531	0569	0607	0645	0682	0719	0755	4	8	11	15	19	23	26	30	34
12	0792	0828	0864	0899	0934	0969	1004	1038	1072	1106	3	7	10	14	17	21	24	28	31
13	1139	1173	1206	1239	1271	1303	1335	1367	1399	1430	3	6	10	13	16	19	23	26	29
14	1461	1492	1523	1553	1584	1614	1644	1673	1703	1732	3	6	9	12	15	18	21	24	27
15	1761	1790	1818	1847	1875	1903	1931	1959	1987	2014	3	6	8	11	14	17	20	22	25
16	2041	2068	2095	2122	2148	2175	2201	2227	2253	2279	3	5	8	11	13	16	18	21	24
17	2304	2330	2355	2380	2405	2430	2455	2480	2504	2529	2	5	7	10	12	15	17	20	22
18	2553	2577	2601	2625	2648	2672	2695	2718	2742	2765	2	5	7	9	12	14	16	19	21
19	2788	2810	2833	2856	2878	2900	2923	2945	2967	2989	2	4	7	9	11	13	16	18	20
20	3010	3032	3054	3075	3096	3118	3139	3160	3181	3201	2	4	6	8	11	13	15	17	19
21	3222	3243	3263	3284	3304	3324	3345	3365	3385	3404	2	4	6	8	10	12	14	16	18
22	3424	3444	3464	3483	3502	3522	3541	3560	3579	3598	2	4	6	8	10	12	14	15	17
23	3617	3636	3655	3674	3692	3711	3729	3747	3766	3784	2	4	6	7	9	11	13	15	17
24	3802	3820	3838	3856	3874	3892	3909	3927	3945	3962	2	4	5	7	9	11	12	14	16
25	3979	3997	4014	4031	4048	4065	4082	4099	4116	4133	2	3	5	7	9	10	12	14	15
26	4150	4166	4183	4200	4216	4232	4249	4265	4281	4298	2	3	5	7	8	10	11	13	15
27	4314	4330	4346	4362	4378	4393	4409	4425	4440	4456	2	3	5	6	8	9	11	13	14
28	4472	4487	4502	4518	4533	4548	4564	4579	4594	4609	2	3	5	6	8	9	11	12	14
29	4624	4639	4654	4669	4683	4698	4713	4728	4742	4757	1	3	4	6	7	9	10	12	13
30	4771	4786	4800	4814	4829	4843	4857	4871	4886	4900	1	3	4	6	7	9	10	11	13
31	4914	4928	4942	4955	4969	4983	4997	5011	5024	5038	1	3	4	6	7	8	10	11	12
32	5051	5065	5079	5092	5105	5119	5132	5145	5159	5172	1	3	4	5	7	8	9	11	12
33	5185	5198	5211	5224	5237	5250	5263	5276	5289	5302	1	3	4	5	6	8	9	10	12
34	5315	5328	5340	5353	5366	5378	5391	5403	5416	5428	1	3	4	5	6	8	9	10	11
35	5441	5453	5465	5478	5490	5502	5514	5527	5539	5551	1	2	4	5	6	7	9	10	11
36	5563	5575	5587	5599	5611	5623	5635	5647	5658	5670	1	2	4	5	6	7	8	10	11
37	5682	5694	5705	5717	5729	5740	5752	5763	5775	5786	1	2	3	5	6	7	8	9	10
38	5798	5809	5821	5832	5843	5855	5866	5877	5888	5899	1	2	3	5	6	7	8	9	10
39	5911	5922	5933	5944	5955	5966	5977	5988	5999	6010	1	2	3	4	5	7	8	9	10
40	6021	6031	6042	6053	6064	6075	6085	6096	6107	6117	1	2	3	4	5	6	8	9	10
41	6128	6138	6149	6160	6170	6180	6191	6201	6212	6222	1	2	3	4	5	6	7	8	9
42	6232	6243	6253	6263	6274	6284	6294	6304	6314	6325	1	2	3	4	5	6	7	8	9
43	6335	6345	6355	6365	6375	6385	6395	6405	6415	6425	1	2	3	4	5	6	7	8	9
44	6435	6444	6454	6464	6474	6484	6493	6503	6513	6522	1	2	3	4	5	6	7	8	9
45	6532	6542	6551	6561	6571	6580	6590	6599	6609	6618	1	2	3	4	5	6	7	8	9
46	6628	6637	6646	6656	6665	6675	6684	6693	6702	6712	1	2	3	4	5	6	7	7	8
47	6721	6730	6739	6749	6758	6767	6776	6785	6794	6803	1	2	3	4	5	5	6	7	8
48	6812	6821	6830	6839	6848	6857	6866	6875	6884	6893	1	2	3	4	4	5	6	7	8
49	6902	6911	6920	6928	6937	6946	6955	6964	6972	6981	1	2	3	4	4	5	6	7	8
50	6990	6998	7007	7016	7024	7033	7042	7050	7059	7067	1	2	3	3	4	5	6	7	8
51	7076	7084	7093	7101	7110	7118	7126	7135	7143	7152	1	2	3	3	4	5	6	7	8
52	7160	7168	7177	7185	7193	7202	7210	7218	7226	7235	1	2	2	3	4	5	6	7	7
53	7243	7251	7259	7267	7275	7284	7292	7300	7308	7316	1	2	2	3	4	5	6	6	7
54	7324	7332	7340	7348	7356	7364	7372	7380	7388	7396	1	2	2	3	4	5	6	6	7

LEFAX,® (Marca Reg.) Philadelphia 7, Pa. **Reprinted Nov., 1961**

Figure 62. Logarithms of numbers. (Courtesy of Lefax)

	0	1	2	3	4	5	6	7	8	9	Proportional Parts								
											1	2	3	4	5	6	7	8	9
55	7404	7412	7419	7427	7435	7443	7451	7459	7466	7474	1	2	2	3	4	5	5	6	7
56	7482	7490	7497	7505	7513	7520	7528	7536	7543	7551	1	2	2	3	4	5	5	6	7
57	7559	7566	7574	7582	7589	7597	7604	7612	7619	7627	1	2	2	3	4	5	5	6	7
58	7634	7642	7649	7657	7664	7672	7679	7686	7694	7701	1	1	2	3	4	4	5	6	7
59	7709	7716	7723	7731	7738	7745	7752	7760	7767	7774	1	1	2	3	4	4	5	6	7
60	7782	7789	7796	7803	7810	7818	7825	7832	7839	7846	1	1	2	3	4	4	5	6	6
61	7853	7860	7868	7875	7882	7889	7896	7903	7910	7917	1	1	2	3	4	4	5	6	6
62	7924	7931	7938	7945	7952	7959	7966	7973	7980	7987	1	1	2	3	3	4	5	6	6
63	7993	8000	8007	8014	8021	8028	8035	8041	8048	8055	1	1	2	3	3	4	5	5	6
64	8062	8069	8075	8082	8089	8096	8102	8109	8116	8122	1	1	2	3	3	4	5	5	6
65	8129	8136	8142	8149	8156	8162	8169	8176	8182	8189	1	1	2	3	3	4	5	5	6
66	8195	8202	8209	8215	8222	8228	8235	8241	8248	8254	1	1	2	3	3	4	5	5	6
67	8261	8267	8274	8280	8287	8293	8299	8306	8312	8319	1	1	2	3	3	4	5	5	6
68	8325	8331	8338	8344	8351	8357	8363	8370	8376	8382	1	1	2	3	3	4	4	5	6
69	8388	8395	8401	8407	8414	8420	8426	8432	8439	8445	1	1	2	2	3	4	4	5	6
70	8451	8457	8463	8470	8476	8482	8488	8494	8500	8506	1	1	2	2	3	4	4	5	6
71	8513	8519	8525	8531	8537	8543	8549	8555	8561	8567	1	1	2	2	3	4	4	5	5
72	8573	8579	8585	8591	8597	8603	8609	8615	8621	8627	1	1	2	2	3	4	4	5	5
73	8633	8639	8645	8651	8657	8663	8669	8675	8681	8686	1	1	2	2	3	4	4	5	5
74	8692	8698	8704	8710	8716	8722	8727	8733	8739	8745	1	1	2	2	3	4	4	5	5
75	8751	8756	8762	8768	8774	8779	8785	8791	8797	8802	1	1	2	2	3	3	4	5	5
76	8808	8814	8820	8825	8831	8837	8842	8848	8854	8859	1	1	2	2	3	3	4	5	5
77	8865	8871	8876	8882	8887	8893	8899	8904	8910	8915	1	1	2	2	3	3	4	4	5
78	8921	8927	8932	8938	8943	8949	8954	8960	8965	8971	1	1	2	2	3	3	4	4	5
79	8976	8982	8987	8993	8998	9004	9009	9015	9020	9025	1	1	2	2	3	3	4	4	5
80	9031	9036	9042	9047	9053	9058	9063	9069	9074	9079	1	1	2	2	3	3	4	4	5
81	9085	9090	9096	9101	9106	9112	9117	9122	9128	9133	1	1	2	2	3	3	4	4	5
82	9138	9143	9149	9154	9159	9165	9170	9175	9180	9186	1	1	2	2	3	3	4	4	5
83	9191	9196	9201	9206	9212	9217	9222	9227	9232	9238	1	1	2	2	3	3	4	4	5
84	9243	9248	9253	9258	9263	9269	9274	9279	9284	9289	1	1	2	2	3	3	4	4	5
85	9294	9299	9304	9309	9315	9320	9325	9330	9335	9340	1	1	2	2	3	3	4	4	5
86	9345	9350	9355	9360	9365	9370	9375	9380	9385	9390	1	1	2	2	3	3	4	4	5
87	9395	9400	9405	9410	9415	9420	9425	9430	9435	9440	0	1	1	2	2	3	3	4	4
88	9445	9450	9455	9460	9465	9469	9474	9479	9484	9489	0	1	1	2	2	3	3	4	4
89	9494	9499	9504	9509	9513	9518	9523	9528	9533	9538	0	1	1	2	2	3	3	4	4
90	9542	9547	9552	9557	9562	9566	9571	9576	9581	9586	0	1	1	2	2	3	3	4	4
91	9590	9595	9600	9605	9609	9614	9619	9624	9628	9633	0	1	1	2	2	3	3	4	4
92	9638	9643	9647	9652	9657	9661	9666	9671	9675	9680	0	1	1	2	2	3	3	4	4
93	9685	9689	9694	9699	9703	9708	9713	9717	9722	9727	0	1	1	2	2	3	3	4	4
94	9731	9736	9741	9745	9750	9754	9759	9763	9768	9773	0	1	1	2	2	3	3	4	4
95	9777	9782	9786	9791	9795	9800	9805	9809	9814	9818	0	1	1	2	2	3	3	4	4
96	9823	9827	9832	9836	9841	9845	9850	9854	9859	9863	0	1	1	2	2	3	3	4	4
97	9868	9872	9877	9881	9886	9890	9894	9899	9903	9908	0	1	1	2	2	3	3	4	4
98	9912	9917	9921	9926	9930	9934	9939	9943	9948	9952	0	1	1	2	2	3	3	4	4
99	9956	9961	9965	9969	9974	9978	9983	9987	9991	9996	0	1	1	2	2	3	3	3	4

Figure 62. Continued

not have to find out the difference between successive numbers in the table, but merely read the proportional parts.

If you are finding the number when you know the logarithm, the process is just the opposite; you find the number past the one you are in the table, then look in the proportional parts column for the correct column. If you have a logarithm as 2.2746, the mantissa 2746 would be just past a value of 2742 shown for 188; a difference of 4. Looking to the right, there is no 4 in the proportional parts, so you use the nearest figure, 5. Reading the proportional part at the top of the column, you find 2, so the number you want is 1882. The characteristic of 2 of the antilogarithm means you have 3 digits to the left of the decimal point, so the antilogarithm is 188.2.

LOGARITHMS OF FUNCTIONS

Logarithmic sines and cosines are shown in Figure 60 and in the Appendix. Since they are read directly from angles, without first determining the number, there is no way to determine the characteristic except by reference to the tables. Also, these functions are ratios less than 1, so the characteristics are minus numbers. Rather than use negative numbers, common practice is to add 10 to all logarithmic functions and show −10 at the end. For example, the natural sine of 5° is 0.0872. From log tables (Figure 62), the log of 0.0872 is −2.9405, the logarithmic sine. But looking at the logarithmic tables for sine of 5° (Figure 60), we have 8.9403, that is, −2.9405 = 8.9403 − 10. The difference is due to the fact that 0.0872 is correct only to 3 digits, but the logarithmic sine is correct to 4.

Using the logarithmic functions, the solution of the problem in Figure 59 is as follows.

$$d = \frac{50}{\cos 25^\circ} \qquad b = d \sin 25^\circ$$

$$\begin{aligned}
\log 50 &= 1.6990 \\
-\log \cos 25^\circ &= 9.9573 \\
\text{antilog } 55.16 &= \overline{1.7417} \\
+\log \sin 25^\circ &= 9.6259 \\
\text{antilog } 23.31 &= \overline{1.3676} \quad (-10\text{'s omitted})
\end{aligned}$$

$$b = 23.31$$
$$d = 55.16$$

In summary, remember:

1. Tabular values for logarithms of *numbers* are the mantissas only.

2. Tabular values for logarithms of *functions* are characteristics and mantissas, but with −10 omitted.

Logarithms for functions which are over 1 follow the same convention, resulting in tabular values such as 10.3356, meaning 0.3356. This occurs not for the sine or cosine, but for tangents of angles over 45°.

Since the sine of an angle is the same as the cosine of the other angle in a right triangle (sine $x = \cos(90\text{-}x)$), tables read both from top and bottom. If sine values are given, cosines can be found from the same table, reading in the opposite direction and for the other angle of the triangle.

The −10 of logarithmic functions may be omitted from calculations and added to the result, or subtracted, as will give a reasonable answer. If an error is made which will result in a wrong answer by a factor of 10,000,000, which results from a failure to subtract 10 from the logarithm, it can hardly be overlooked!

To multiply numbers, add their logarithms; to divide subtract the logarithms. Thus

$$\text{for side } d = \frac{50}{\cos 25°}$$

or

$$\log d = \log 50 - \log \cos 25°$$

we have

$$\log 50 = 1.6990$$

subtract

$$\log \cos 25° = 9.9573 - 10$$

In order to subtract 9 from 1, show that 1.6990 can be written 11.6990 − 10.

Thus

$$\begin{array}{r} 11.6990 - 10 \\ -(\ 9.9573 - 10) \\ \hline \end{array} \qquad \text{or} \qquad \begin{array}{r} 11.6990 \\ +\ 9.9573 \\ \hline 1.6990 = 55.16 \end{array}$$

The four-place functions and logarithms shown in this chapter are for practice use and illustrate the principles involved. For field use, at least five-place tables are recommended.

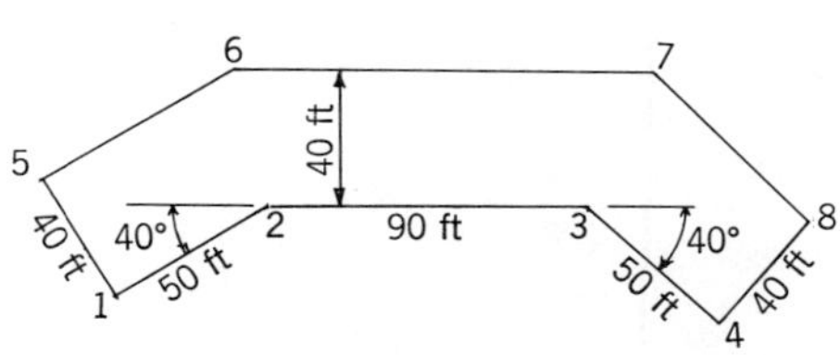

Figure 63. Field Exercise Number 12.

FIELD EXERCISE NUMBER 12

This exercise requires computation of the layout distances in advance of field work. The field work can be completed in about 1½ hours. Each party is to compute and lay out a building as shown in Figure 63.

The entire building may be laid out from four setups by computing the diagonal distances 2–6 and 3–7. Setups are 1, 2, 3, and 4 turning angles and measuring back corners in from each of these points.

Check closure of angle by setup on 5 and 8; measure the actual angle, and record this in your notebook.

Check closure of distance by measuring back building lines 5–6, 6–7, and 7–8, and compare them with calculated dimensions. Show actual measurements in your field book. Each student is to prepare complete field notes for the layout done by his party. This may be done by two-man parties.

Chapter 11

TRAVERSE CLOSURES

A *traverse* is a series of survey lines, which is open or closed, depending on whether it returns to the starting point. If you start from one corner and lay out three sides of a building, you make an *open traverse*. If there is an error in one side or in an angle, you have not found the error. If you measure all four sides and all four angles, you have a *closed traverse*. Any error in the work will appear as a discrepancy in the last angle or distance, as compared with the plan, and this difference between what the dimensions are and what they should be is the *error of closure*. The sides of the building are *courses*, and the direction of these sides are *bearings*. A bearing indicates the direction of a line, as does an azimuth, in degrees; the difference is that bearings are measured to east or west of north or south lines, always resulting in an angle less than 90°, to make figuring easier. Which lines are the north or south bearings depends on the starting point.

If a building is setting north and south and has square corners, one who walks along the four sides would go north, east, south, and west; these are the *bearings* of the wall lines.

The general contractor or developer may lay out the streets in a project, and there may be building locations at oblique angles to each other and to the streets. These dimensions are often not given completely on the plans, particularly if the property is all under one ownership. On a rectangular lot, the layout engineer may check the layout by measuring to a back lot line; on all other types of layout, a computed closure that is adapted to the angles actually used is necessary. *The purpose of computing traverse closures is to check the accuracy of the field work.* This method is used to check traverses made to establish points for use in topographical mapping, and is in constant use in land surveying. Such problems are an important part of examinations for registered land surveyors. The nomenclature of traverses, the standards of accuracy, and the method of computation show but minor variations.

Nearly any problem involving determination of distances and angles, part of which are determined by surveying methods, can be worked out by traverse closures; in this manual, the method is confined to use for checking purposes.

NATURAL FUNCTIONS VS. LOGARITHMIC FUNCTIONS

Traverse closures can be made either by natural or by logarithmic functions. The use of natural functions is confined largely to offices where a calculator is available, and to older surveyors or to those who have never had the opportunity to learn logarithms. A man who has considerable experience with arithmetical computations can figure these closures as fast as one less skilled can with logarithms, and the time actually used in multiplication and division is so small that there may not be a significant advantage in the faster method.

In this manual, logarithmic functions are used, since men with little arithmetical experience and no equipment can work most efficiently with logarithms. You can readily change to natural functions later if desired. For computation of small angular offsets and short distances, you will probably use natural functions.

LATITUDES AND DEPARTURES

To close a traverse, add up all the distances of a traverse, and use their north-south and east-west projections, to see if they are correct. If you have a building 10 ft × 10 ft, you know the distances on one side must add up to the distances on the other. This is a traverse closure.

If the courses are at an oblique angle (any direction except due north, south, east, or west), you must determine their north and east projections. For example, for each 100 ft that you travel in a direction 28°17′ east of north, you travel 88.06 ft further north, as shown in Figure 64. This is called the *latitude,* because of the fact that it is actually a change in latitude; it is equal to the length of the course, times the cosine of the bearing. At the same time, you go easterly, or depart, 47.38 ft east of the north-south line, or line of longitude, you started from. The east-west movement is the *departure*. The term is from land surveying methods in which the surveyor, using a compass, ran a north or south line to the land corner which had been set in an east-west direction. Because of the earth's curvature, these lines did not hit on the points the same distance east or west from a base line, but missed the measured points. They were said to *depart,* in terms of plane surveying, from the line they would have hit on a flat surface. *Working latitudes and departures* is an old expres-

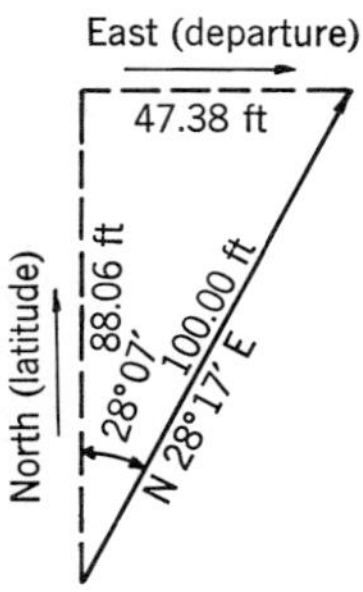

Figure 64. Latitude and departure.

sion meaning to figure traverse closures. In the military, and for some other uses, *x and y coordinates* are terms applied to latitude and departure.

BEARING COMPUTATIONS

If you run a traverse which returns to the starting point, the total of all northerly latitudes must equal the southerly ones, and the easterly departures must equal the westerly ones. The traverse calculations in Figure 65 illustrate the form used to add up these latitudes and departures. In this illustration, the bearings are taken clockwise around the circuit; they could be counterclockwise, and some engineers follow this practice. Corner angles inside the figure are *interior angles.*

Deflection angles are changes of direction of courses, or the difference between the interior angles and 180°. They are right or left, as shown in Figure 65. Deflection angles are used to show changes in alignment for highways. They are right or left, depending on the direction the line is run; for example, the deflection angle of *65° right* labeled in the figure is right because the bearings are being figured in the clockwise direction. If the bearings were considered in the opposite direction, the deflection angle would be *65° left.*

THE CALCULATION

The work is done in the following order, as illustrated in Figure 65.

1. Draw a sketch of the traverse, showing approximately the relationship of distances and bearings.

2. Label the sketch with all angles. Usually the angles in the notes are interior angles, because deflection angles cannot be as accurately measured. If deflection angles have been figured in the field book they should be checked anyway; if there is an error in the angles, check first for errors in the field book.

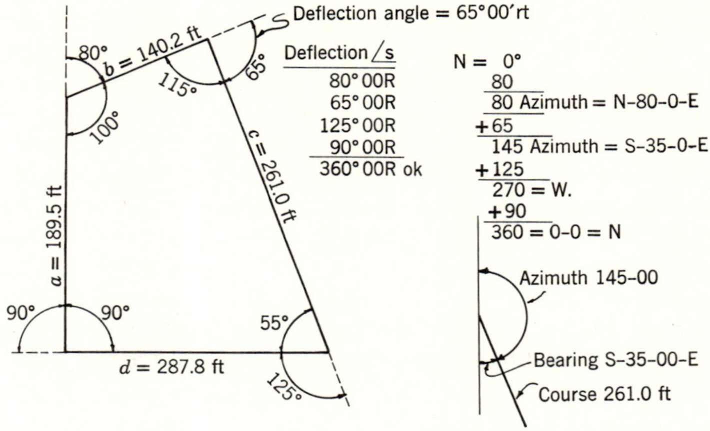

Figure 65. Illustrative example of traverse closure.

Add up all angles to be sure there is no error, or to distribute error where it probably occurred. *Interior angles should add up to a multiple of 180°; deflection angles should add up algebraically to 360°*. That is, all the right deflections must be greater than the sum of the left deflections by 360° if traverse lines are calculated clockwise, because the line has made a complete circuit. If the angles do not add up correctly and no correction is made, the error will show up when you compute the starting bearing from the bearing of the last course, and will appear in the last angle.

3. Select a north direction. In some cases, this is established. In most cases, you may call any convenient direction, north. You should select a course as due north or east-west, whether you really have a course near such a direction or not, to save a calculation. If two courses are at right angles to each other, using one of them for the north direction will save computing two sets of bearings. Likewise, if two courses are parallel, they should be used for one of the directions for which a calculation is not required, such as south or east. A traverse closure may be completely checked by computing it once with one course as north, then a second time using a different course as north. All bearings and functions will then be different, but the error of closure will be the same.

4. Calculate all course bearings. In this example, the 189.5′ course is chosen as north. The first deflection angle, 80-00 right,is the same direction in which azimuths are measured, so this is added to zero to give an azimuth of 80-00 for the course 140.2′. The next deflection angle, 65-00 right, is added to the previous azimuth, giving 145-00 azimuth for the 261.0

course. As shown on the sketch below, this is equivalent to 180-145 = S 35-00 E. Each of the azimuths is calculated around the circuit, and each converted to bearings. The last calculation yields 360-00, which is the same as 00-00, indicating that the calculation of azimuths has been correct. It does not, however, check the bearings.

5. Enter course distances and bearings in the form provided. Using the directions on the bearings for each course, draw lines in the vacant spaces which will not be used; for example, if the bearing is N 80-00 E, draw a line in the S and W spaces to make sure that no figures will be written erroneously in those spaces. In this form, east and west are shown first, as most students think of sine before cosine, and there should be a definite arrangement to avoid the possibility that the latitudes and departures will be accidentally reversed. Since there is no check on this closure except to repeat the entire process, it is important that every possible error be avoided. Remember the description "latitude and departure" or "northerly and easterly" is *opposite* to the way you read the functions "sine and cosine." The latitude, or north direction, corresponds to the *cosine,* not the sine.

6. Tabulate vertically all distances and bearings as shown below table of Figure 65. By writing these all at once, and looking them up at one time, much time is saved and an error is less likely. Notice that the course distances go into the middle of the sets of calculations below, and the sine and cosine corresponding to that course distance immediately above and below it. You then add up and down from this distance, so you need write the distance and its logarithm but once. Also, the sine products will all be at the top and the cosine products at the bottom, making confusion between the two less likely. The sines are written all on the same line, so they will not be confused with the cosines.

7. Enter the tables for logarithms of numbers and for logarithms of the distances, then tables of logarithmic sines and cosines for the functions.

8. Add logarithmic distances to sines upward, and to logarithmic cosines downward.

9. From these results, find values of latitudes and departures and write in lower computations next to their logarithms.

10. Enter these computed latitudes and departures in the form and add them. Any difference between north and south, or between east and west, is error of closure in that direction.

CLASSROOM EXERCISE NUMBER 4

Each student will sketch the oblique building that was laid out in Field Exercise Number 12, with the left-hand end wall as north, and complete traverse closure using the form and method shown in Figure 65. This may be done either on a sheet of paper or in the notebook, at the instructor's option. If four place tables are used, closure error should be not over 0.1′. Five place tables should give closure with the same error as original layout computations.

Use actual measured distances and angles from your field problem. You will have to figure the missing angles at the back corners, because not all the angles were measured in this exercise.

Chapter 12

PEG LEVEL ADJUSTMENT

If a transit, engineer's level, or builder's level is adjusted so that the bubble in the vial is centered when the line of sight is level, the instrument will accurately read all levels, regardless of other adjustments which may be required.

To do this, the telescope is first placed in a line known to be level, and while it is in this position the bubble tube is raised or lowered at one end by an adjusting screw until the bubble is centered. The telescope is then in adjustment. As with a carpenter's level, the principle is the same for any instrument using a spirit level. If you block the ends of a carpenter's level until the bubble is centered, and then reverse the level on the blocks, the bubble will remain centered only if the spirit level is reading true. The telescope level uses the same principle, differing only in that the carpenter's level may be adjusted by moving the supports to a level position and adjusting the bubble to the supports; for the telescopic level, marks on the level rod, rather than the pegs in the ground, are determined as level with each other, and the telescope bubble tube adjusted to these marks.

DETERMINATION OF A LEVEL LINE

First, you may establish a level line by determining the relative elevation of two pegs in the ground. This is the *peg method*. You can do this even though the instrument is not in adjustment, using the principle that slope of the line of sight of the telescope is always the same when the bubble is centered, even though the line of sight is not level. If you determine the elevation difference of two points by setting up near one of them, and then determine the elevation difference from a setup near the other, the two differences will be the same only if the instrument is in adjustment. If it is not in adjustment, the average of the two elevation differences will be the correct difference.

Figure 66. Initial readings.

Two pegs are driven as far apart as possible, within 300 ft, Figure 66, and the instrument set up as close as possible to one of them; usually this is about 8 ft, as the telescope will not focus at a very short distance. This step is illustrated in Figure 66, with the method of recording the rod readings. These readings show that the pegs (stakes) have a difference in elevation of 0.65 ft. At this point we do not know if the instrument is in adjustment, and therefore do not know if the difference of 0.65 ft. is correct.

The next setup, as shown in Figure 67, is near stake *B*, from which setup the two rods are again read. It is important to label the shots as shown, and keep the rod readings at the second setup on the same line as the labels for them are on. You now have a second difference between the peg elevations, or 0.57 ft, indicating that the instrument is out of adjustment. If these differences were within 0.01 ft of each other, you would consider the instrument to be in adjustment and go no further in this process.

The true difference between the elevation of the points is the average of 0.57 ft and 0.65 ft. To find this, you add $0.57 + 0.65 = 1.22$ ft, and divide by $2 = 0.61$ ft. This is the correct difference, so if the instrument is set to read this difference it will be in adjustment.

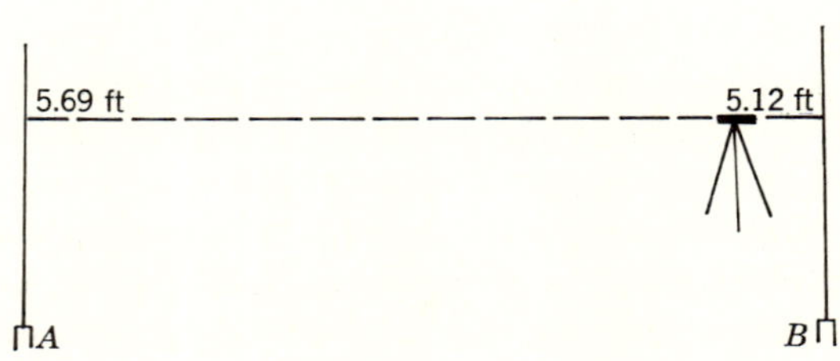

Figure 67. Check reading.

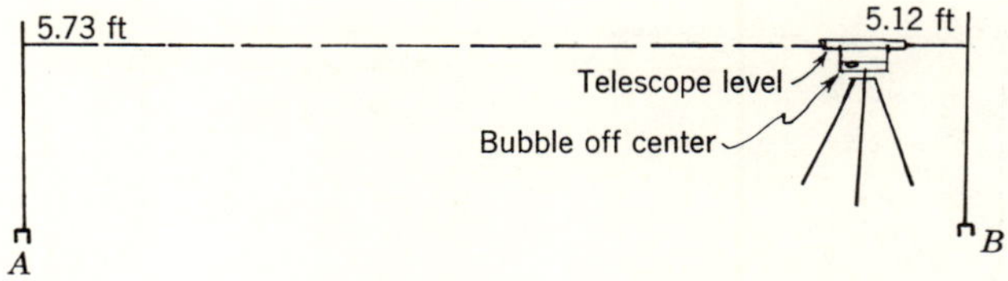

Figure 68. Setting the telescope level.

SETTING THE TELESCOPE LEVEL

Since you are much closer to one point than to the other, the error in reading the closer rod is much less (Figure 68). You may assume the error on the near rod is zero; that is, that the reading of the near rod is correct. If you are set up near rod *B*, you show in another column the reading on rod *B* as the correct one, here 5.12 ft. The true difference is 0.61 ft, so this is next written in the same column. Now if the instrument is to read a difference of 0.61 ft, it must read 0.61 + 5.12 = 5.73 ft on the far rod. So you set the instrument on 5.73 ft on the *far* rod, using the leveling head screws and allowing the bubble to go where it will. The telescope is now level.

Do not attempt to adjust the telescope while it is pointed toward the nearest rod, at any time. Also, be sure you apply the error to the *near rod reading.*

The 0.61 ft is added to 5.12 ft, rather than deducted, because that is the way the difference appears in the column under setup *B*. If the rod *A* reading was less than the rod *B* reading, then the difference would be subtracted from the rod *B* reading.

ADJUSTING THE BUBBLE TUBE

Now that the telescope is level (Figure 69), lower or raise one end of the bubble tube, without changing the pointing of the telescope, until the bubble is centered. To prevent changing the pointing of the telescope during this process, be sure it is securely clamped. In most instruments, this adjustment is made by *double capstan nuts;* these are nuts with holes in the sides, such as in a ship's capstan, into which a brass pin is inserted to turn the nut. The nuts are made in this way so that no one will be able to turn them with a wrench or screwdriver, which could easily strip the threads. There is a nut above and below the flat strip that holds the end of the bubble tube, and the nut toward which the tube end is to move must be loosened first in order to allow movement of the second nut.

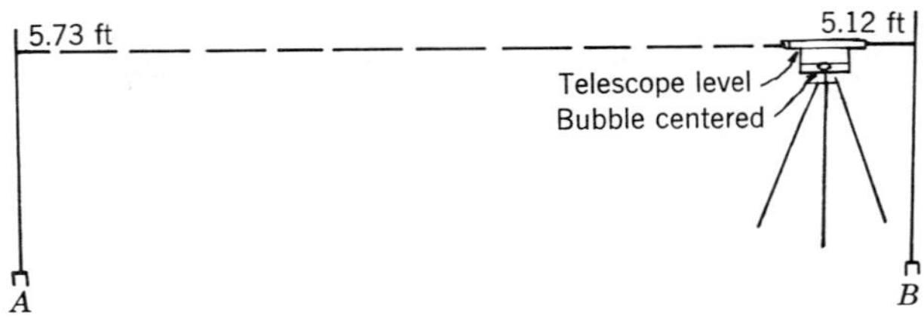

Figure 69. Adjusting the bubble tube.

After adjusting the tube, check the reading on the rod, on the far point again; if it has moved, reset the telescope on the correct reading and re-adjust the bubble tube.

FINAL CHECK READING

To be sure no error has been made in the adjustment, again move to a point near the opposite rod—in this case the A rod, and set up again. Read both rods; if you have made the adjustment properly, the difference between the rod readings should still be 0.61 ft (Figure 70).

If this final check reading is more than 0.01 ft from the previous check reading, an error has been made. Some common causes of discrepancies are listed here.

1. The original error was very large, such as over 0.10 ft in 200 ft. The assumption that the near rod reading was initially correct may have been in error by more than 0.01 ft.
2. Adjustment may have been made on the near rod, or the correction applied to the far rod rather than vice versa.
3. The leveling screws on the instrument may have been loose, allowing the telescope to rock back and forth, or you have left them loose after adjustment.

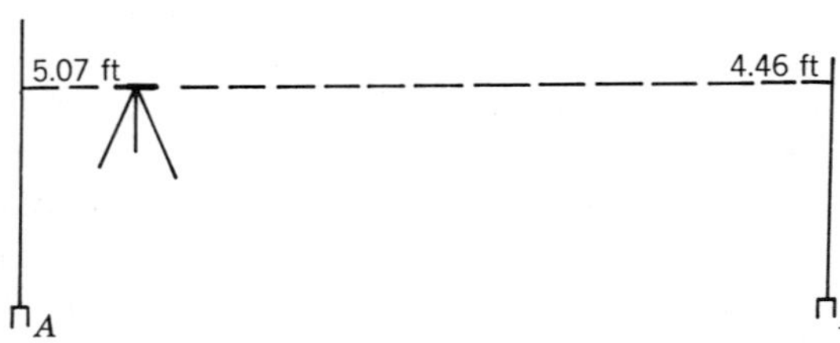

Figure 70. Final check reading.

4. The capstan screws may not have been fully tightened, allowing movement of the tube; or the tube itself may be loose in its metal supports.

5. A reading was taken when the bubble was not centered.

6. You may have an instrument that is not capable of being adjusted as closely as 0.01 ft in 100 ft. This may be checked in Chapter 9.

In any case except the last, you use the difference between the last readings, and average it with the preceding difference as you did before; adjust the instrument at location *B,* and check again at *A*. If the instrument error was off a large amount initially, it may take several trials to adjust it.

If the difference in elevation between the two points used is less than the error of the instrument, the initial elevation differences will be in opposite directions for the two points, depending on the location of the setup. The simplest way out of this situation is to reset the points, being sure there is a greater difference between them. If two differences opposite in direction are averaged, the average is half of one difference minus the other.

FIELD EXERCISE NUMBER 13

The instructor may attach rods to trees or use rod faces on fixed objects to avoid use of rodmen. Each student is given an instrument, which has been slightly thrown out of adjustment, to adjust the bubble tube. He has an adjusting pin, if available.

When the student completes adjustment and checking of his own instrument, he exchanges it with another student for an instrument which has been adjusted; in this way, no instrument is put away without having been checked by two people. If the two differ on whether the instrument is in adjustment, the last person to check it puts a note in the box, and it is set aside for the instructor to check later.

Each student should check two instruments in 2½ hours; there are a few people who are unable to understand or to properly manipulate the instruments for this exercise. Consequently, some persons, in order to complete the work, must take an additional period, depending on the standards of the instructor.

A common student error when making the calculation for this exercise is to determine the doubled error of the instrument, divide by two, and add or subtract this amount from the far sight rather than to use the difference between stake elevations as described in this chapter. If correctly done, the results of such a method are the same as if the stake difference is used. However, determination of whether the error is to be added or subtracted requires more understanding of the work than does the method outlined above, and the chance of correct applications by students is about 50%. The method outlined is recommended because an error is large enough to be evident. If the instrument error is a few hundredths, and the calculation shows the correction to be several tenths, as it will if you apply the peg difference the wrong direction, it is evident that you have made an error. Therefore, do not figure the exact instrument error or write it down, unless you know exactly what you are doing; the procedure is not recommended for first-semester students.

Chapter 13

EXCAVATION VOLUME

Both contractors and engineers calculate volume of excavation for estimating cost, checking job costs as the work proceeds, and for final payment. Often the contractor, when bidding a job for which he is to be paid on a per yard basis, will check the engineer's measurements of calculations for an error. Such an error gives the contractor an opportunity to bid "off balance"; by reducing his unit price on the quantities which will be large, and increasing it on the quantities which will be small, he can keep his initial bid price below his competitor's and still collect more in the final payment.

VOLUME OF SOLIDS

Figure 71 shows the three simple solids—the square box, with a volume equal to the product of its sides, or the area of one end times the length. If the ends are the same size and shape, the volume will be the end area times the length, regardless of the shape of the ends. This may be a cylinder, having circles for ends, or a prism with triangles for ends.

The volume of the cone is one-third the product of the end area and the length, regardless of the shape of the end area (or base). The volume of a wedge, which differs from the cone only in that the area reduces throughout its width at the same time, rather than reducing from one end to the other in *both* directions at once, is the end area times the length, times one-half. A wedge is actually one-half of the square box, cut in two along one diagonal. If the base of the wedge is other than a rectangle, such as a circle or triangle, you can not cut a wedge from it that would be a straight line at one end. You would cut a wedge that would follow this rule for calculating volume only if you cut the original solid in half diagonally, which would give it an unusual shape.

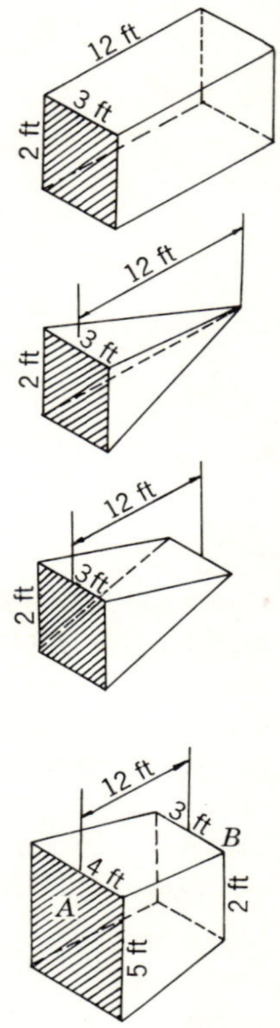

Figure 71. How to figure excavation.

AVERAGE END AREA

How would you calculate the volume of the last solid in Figure 71? It can be cut up into boxes, wedges, and cones, but this is a lot of work, and earth is not expensive enough to justify this work. If you average the end areas and multiply by the length, the resulting volume will be correct for the box and the wedge contained in this figure (this rule is accurate for boxes and wedges). The cones will be overstated; that is, you will have used half the product of the ends times the length, instead of one-third, as you should for a cone. This gives you an answer which is too large, but is in most cases an acceptable error.

Figure 72 shows two cross-sections, or slices, taken across a roadway at *A* and at *B*. By drawing the present grade and then the grade you want to have on a sheet of paper to the same scale, you can calculate the area of each of these end sections in square feet, either by cutting it up into triangles and trapezoid to figure by arithmetic, or by a calculator. Point *A*

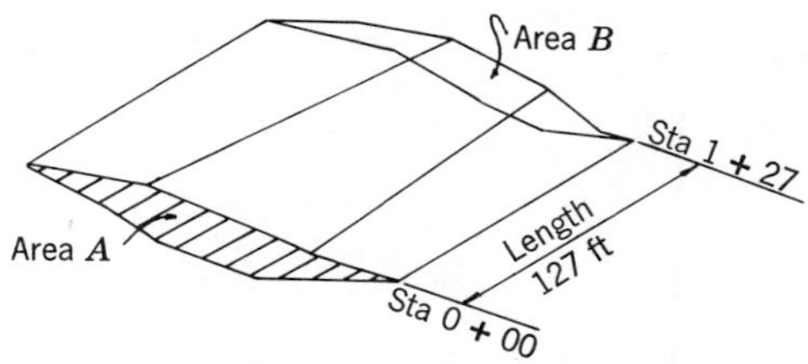

Figure 72. Actual computation by average end area. Volume $= \frac{(\text{area } A + \text{area } B) \times \text{length.}}{2}$

is station 0 + 00 on a centerline, and B is station 1 + 27, or 127 ft further. Using the conventional average end area method of calculation, you calculate areas A and B, and multiply the average of the two by the length. Occasionally the *double end area* is used, which means that you multiply the total area by the length, and add up all the stations before dividing by two. The result is the same in either case.

CALCULATION OF CROSS SECTIONS

The area of each cross section, or *end area,* may be calculated by arithmetic, either manually or by calculator, or by a *planimeter,* a machine which calculates the area by running a pointer around the boundary. Because the planimeter is awkward to use and is not exact anyway, arithmetical methods are usually used.

One simple method of arithmetical calculation is to divide the section into trapezoids and triangles by drawing vertical lines at changes of grade; that is, at crooked places in either the original or final grade line. The area of each of these small areas is the width times the average height—half the sum of the height on both sides, whether it be a triangle or trapezoid. As shown in Figure 73 it is an approximation used to measure the height of the area at the estimated midpoint, usually sufficient and much faster than adding the heights and dividing by two.

CROSS SECTION BY CALCULATOR

The arithmetical method described above may be done on an office calculator, but it still requires that each height of area be measured or calculated. By using the rod readings and distances directly as taken in the field for figuring the areas, all intermediate steps may be eliminated.

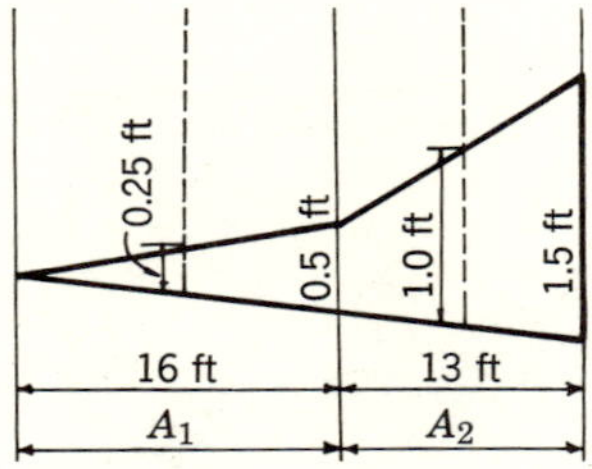

Figure 73. Cross-section areas.

Since the area is plotted for purposes of illustration anyway, the saving in time is not great. However, calculating areas in this way is more accurate, and gives an answer exactly in accordance with the readings taken in the field. For this reason, the calculator is used for measurement of quantities that serve as the basis for payment of the contractor.

A general proof of the calculator method is complicated, but it can be demonstrated that the method is correct in a simple case, such as Figure 74, and in the more general case, Figure 75. Points on these figures are designated by elevation and distance from a base line; $\frac{2.0}{4}$, for example, designates 2.0 ft elevation and 4 ft horizontally from reference lines. In Figure 74, starting at point *A*, the rod reading 0.0 is punched on the keyboard, and multiplied by the horizontal distance to the base line of the preceding point, *C*, which is also 0; the rod reading at *A* is then multiplied by the horizontal distance (4) from the base line of the next point, (*B*), and this is deducted. This process is repeated for *B*, leaving all products on the machine. These products, at *B*, will all be zero. When point *C* is reached, the product of the rod reading at *C*, 5.0 is multiplied by the base distance (4) from the preceding point *B*; this same 5.0 is multiplied by the base distance, 0, of the next point forward on the circuit (*A*), resulting in another zero. The only reading remaining on the ma-

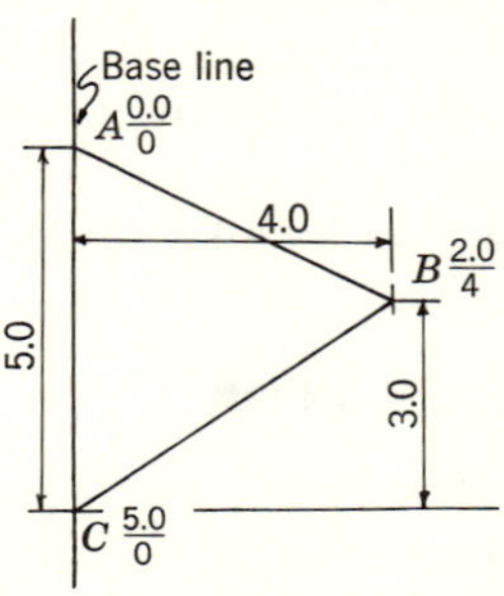

Figure 74. Calculation by calculator.

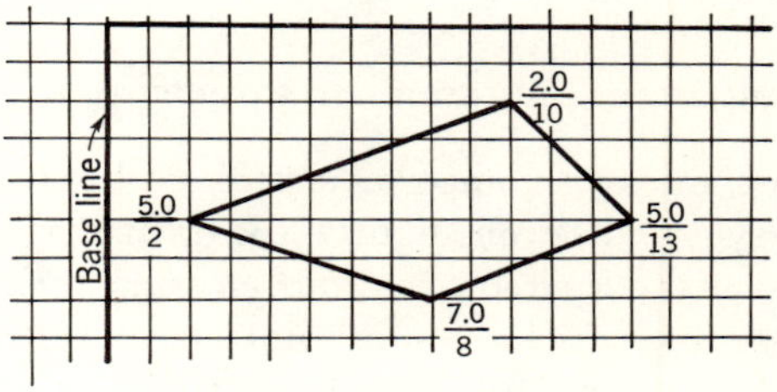

Figure 75. Calculator calculation.

chine is therefore 20, or the double area of the triangle. For this example, it is a complicated way to determine the area of a triangle! This is a general method, however, and is good for any shape. With the same calculations on the machine you could have measured the area of any triangle, knowing only the locations of the points, and without computing the base or altitude.

The chief virtue of this system is that it is suited to simple calculator operation. None of the numerical products of the machine are recorded except the final answer—here, 20 sq ft. This must be divided by two for the actual area. The calculator is equipped to multiply, retain the amount, and to add and subtract new products from the old total, so that the arithmetical operations shown in the examples illustrate what the calculator carries. These figures are not written on paper, unless the calculator prints it on a tape. Many newer machines are printing types.

Figure 75 shows the same method applied to an irregular four-sided figure. Each rod reading is multiplied by the previous distance reading (the bottom part of the "fraction" shown) and from this is deducted the multiplication of the same rod reading by the next (reading clockwise) distance reading. When the circuit is completed, the calculator will read double the area. Either rod readings or elevations may be used, without changing the method.

The sign of the answer (plus or minus) depends on whether the base line is to the right or left, whether rod readings or elevaitons are used, and whether the distance ahead or behind the rod reading is called positive (the other must always be negative). A variation will change the sign, but will not change the numerical value of the answer. If rod readings are being used, if the base line is to the left, and if the distance reading counterclockwise of the rod reading is considered plus, the answer will be plus. Changing two of these will keep the answer plus; changing one or three of them will give you a minus answer. Notice that the actual order of adding up the points is not important, but the direction considered "forward" or "backward" must be consistent, and affects the sign of the answer.

There are three important limitations on this method.

1. The lines must make a complete circuit and return to the same starting point. If they do not, even slightly, the error may be considerable.

2. The lines must make a circuit without crossing themselves. That is, the original and finish grade lines must not cross. When calculating a cross-section area, the cut and fill areas must be taken separately if you want totals and not differences.

3. The base line used must be outside the area. If distances from the base line are right for part of the figure and left for part of it, the answer will be the difference between the two areas. The direction of readings must be reversed at the center, taking each half as a separate area.

For further details on the calculator method, see the distributor of the calculator you have, or the Monroe Co. pamphlet, *Machine Methods for Civil Engineering Calculations.*

ILLUSTRATIVE EXAMPLE

Figure 76 shows the excavation of a building site, using four cross sections; of these, two represent field data, and there are two which are drawn using the new grades with the same field cross sections. The two sections actually taken in the field as station 1 + 27 and 2 + 30, which were either taken at the edges of the new basement excavation, or drawn from a contour map of the site. Station 0 is the north edge of the lot, and represents the end of excavation.

The area of the cross section at station 1 + 27 is calculated back and forward. *Forward* is in the direction of stationing, and *back* is in the opposite direction, toward station 0 + 00. The back section is just out of the hole, and is figured from the natural ground and the finish grade outside the basement hole. The volume between sta 0 and 1 + 27 is then 127 ft times this cross-sectional area times ½—or 127 ft × 330 ft which is the average end area or half the area at station 1 + 27.

The volume from station 1 + 27 to 2 + 30 is obtained by the average end areas, but the end areas for this purpose are only in the hole. The sides of the hole are assumed to be vertical. The section used at station 1 + 27 is the forward-section—the one used to figure the volume forward of this station—and is calculated using the bottom grade of the basement as finish grade. This enables you to use the same original section used for the section out of the hole, and makes it unnecessary to recompute the area at 1 + 27 that is to each side of the hole. You must add up the areas

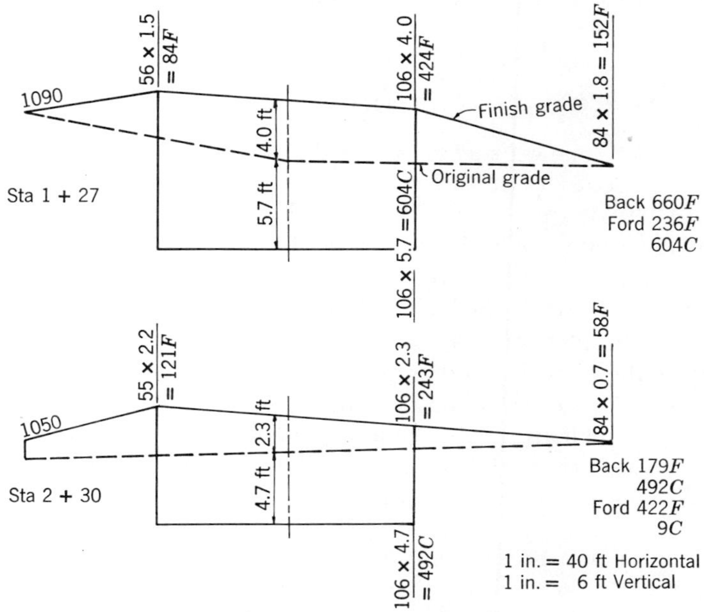

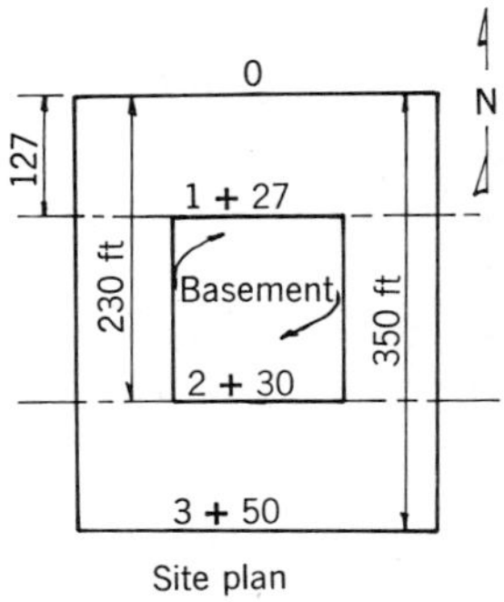

Sta.	Dist.	Fill Area	Fill Av.	Fill Vol.	Cut Area	Cut Av.	Cut Vol.
0		0			0		
	127		330	41,600		0	
1 + 27B		660			0		
1 + 27F		236			604		
	103			21,300		348	57,500
2 + 30B		179			492		
2 + 30F		422			0		
	120		211	25,400		0	
3 + 50		0			0		
		Total		88,300			57,500
				3180	cy		2120 cy

Figure 76. Calculation by end areas—work sheet.

twice to get the total area of the two cross sections at station 1 + 27—once using 604 sf cut for the part in the basement, and once using 424 sf fill for this same portion of the section.

At station 2 + 30, two sections are used, as for station 1 + 27, but here the back section is in the hole, and the forward section, out of the hole, is used to figure the volume from station 2 + 30 to station 3 + 50, where the excavation again drops to zero. *Excavation,* here as in many other places, is used as a general term meaning to move earth. Fill is considered part of excavation, since earth filled one place must be dug somewhere else.

PROCEDURE OF CALCULATION

To make a calculation such as shown in the example, the sections are first plotted on *cross section* or *profile paper*. Cross section paper has a square grid pattern; profile paper has vertical graduation lines further apart and horizontal lines close together. Profile paper is easier to plot elevations on, but does not work out well if the difference in elevation to be plotted is very great, since it is laid out for a few feet per inch scale vertically. When plotting cross sections, the first point is plotted and labeled with the elevation; other points are figured from this one, and each horizontal line of elevation need not be marked. Often sloping cross sections may be high on one side and low on the other, lapping over each other, so that a horizontal line may represent one elevation on one section and another elevation on another section.

When calculating area of cross sections, the work is faster and more accurate if a definite procedure is followed; minimizing the time lost in changing from one scale to another, changing direction of measurement, and computation. The following routine is suggested.

1. Divide the sections into trapezoids and triangles by vertical lines at top and bottom grade breaks, as in Figure 73.

2. Using the horizontal scale, label each vertical line with the distance from the preceding vertical line, as in Figure 76.

3. Using the vertical scale, label each vertical line with the cut or fill distance at the middle of the trapezoid. The middle is estimated by eye.

4. Multiply distances by cuts.

5. Label each area cut or fill.

6. Add the cut areas for a total and write at the end of the section; the same for the fill areas. To avoid confusion between cut and fill quantities

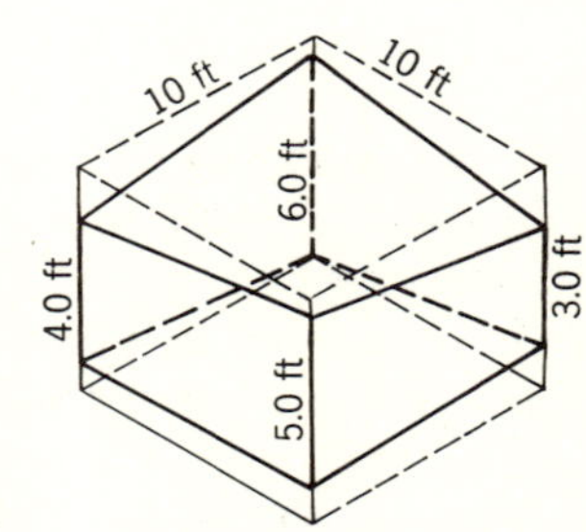

Figure 77. Grid method.

when adding them, it is suggested that the numbers be written in different colors.

Each section is shown as close as possible to its recorded area, so you can easily check the quantity by looking at the section as compared in size with its computed area. Decimal-point errors are greatly reduced by carrying all numbers in the same units, and no accuracy is lost; for example, all horizontal distances to the nearest foot, vertical distances to the nearest tenth, and areas to the nearest square foot. Since only feet are being used until the final answer is obtained, foot marks are not necessary. The mark ′ can easily be mistaken for a 1, so it should not be used.

GRID METHOD

When an area of excavation is large and there are no significant sharp changes of grade, as would occur at creeks or building excavations, the distance between sections and shots along the cross section line can be made the same. It is then unnecessary to plot the sections at all; the shots can all be averaged and multiplied by the ground surface area of the excavation. A part of such a computation is shown in Figure 77. If all four sides are averaged and multiplied by the area of the square (100 sq ft), the volume is obtained just as it would be by plotting and computing the cross sections. The volume you get is the same, but by using such *grids,* the vertical cuts cuts are added up and multiplied once, instead of multiplying them separately and then adding them up. In adding up the cuts or fills, when a grid is incomplete, as at the edge of the field, the dimensions of that grid are reduced proportionately to the area of grid. As in all cross sections, cut and fill areas must be kept separate.

The grid system is well adapted to large, fairly level areas. If applied to small irregular areas, it becomes more work to adapt the grids to the ground than it would be to use cross sections. Also, if you use grids on

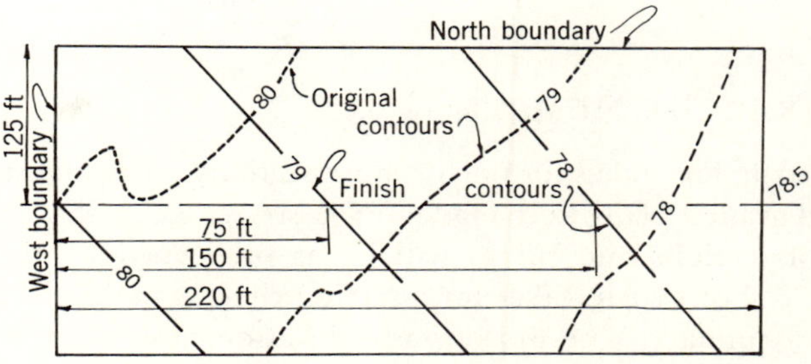

Figure 78. Grade stakes for site grading.

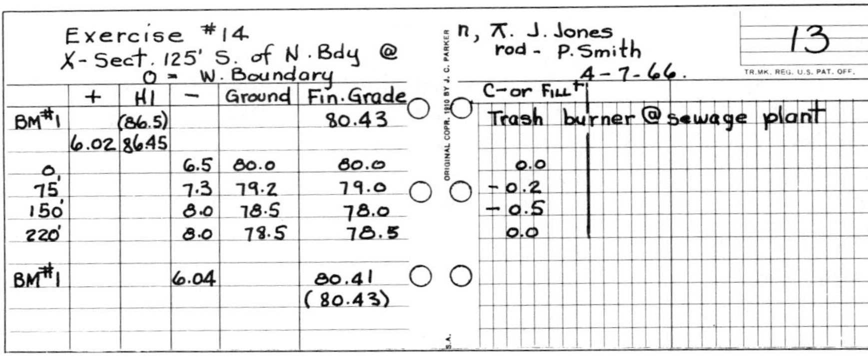

Exercise #14
X-Sect. 125' S. of N. Bdy @
0 = W. Boundary

n, K. J. Jones
rod - P. Smith
4-7-66.

13

TR.MK. REG. U.S. PAT. OFF.

ORIGINAL COPR. 1910 BY J. C. PARKER

	+	HI	−	Ground	Fin. Grade	C− or Fill+
BM#1		(86.5)			80.43	Trash burner @ sewage plant
	6.02	86.45				
0'			6.5	80.0	80.0	0.0
75'			7.3	79.2	79.0	−0.2
150'			8.0	78.5	78.0	−0.5
220'			8.0	78.5	78.5	0.0
BM#1			6.04		80.41 (80.43)	

Figure 78. Continued

50-ft intervals, for example, you must complete the job with this interval throughout. Cross sections may be taken where the ground changes, so one interval between sections may be 50 ft and the next interval several hundred feet. In general, the cross sectional method will give the same accuracy as the grid method with less field work, but often with more office work.

Details of calculation of cross sections may be found in route surveying texts, and details of the grid system in building estimating books. Sometimes long cross sections are called *profiles,* although this term more frequently is used to indicate a cross section taken along the centerline of a highway or other centerline survey. To some extent, the terms profile and cross section are interchangeable.

CLASSROOM EXERCISE NUMBER 5

Using the methods outlined above, compute the quantity of excavation necessary to complete the project designed in Classroom Exercise Number 3.

FIELD EXERCISE NUMBER 14

Set construction stakes for one or more sections across the site for which completed graded plans were made in Classroom Exercise Number 3. The number of sections and stakes will be as prescribed by the instructor.

Figure 78 shows such stakes for a section that was taken 125 ft south of the north boundary. The finish grade elevations are taken from the contour map for the locations as scaled on the map. Original contours are shown, as they would be on the grading plan, but they are not used for layout. A ground shot is taken at each point where a stake is set, and the cut or fill is figured from this shot; if the original contour were used, any error at this point would be reflected in the final grade. For rough grading, the original contours may be used, and a level is not necessary.

These stakes should be set and actually marked, so the students have an opportunity during the course to mark stakes.

Chapter 14

PRACTICAL HINTS

This chapter includes a number of field procedures occasionally needed by surveyors: work under difficult ground conditions, cutting tools, checking adjustment of instruments, setting points in difficult terrain, and setting grades to a sloped, rather than to a level line.

SOFT GROUND

Soft ground and swamp conditions offer special problems to the surveyor, especially in making stable instrument setups. Muck, or peat, often occurs in low areas; this material supports instruments and men very poorly, shrinks almost indefinitely as it may be over 95% water in its original state, and is highly resistant to freezing.

Not only may the instrument itself move in such ground, but the ground may move under the instrument man's weight. The simplest way to make a stable setup in such material is to set the tripod on stakes, as shown in Figure 79. The stakes support the load from below the area disturbed by the operator, and they should be as large as can be conveniently driven. A commonly used size is 2 × 4's, and they may be 3 ft long or more. The resistance of soft material to driving is surprisingly small; a half-inch steel rod can readily be forced by hand pressure to 20 or 30 ft, and piling will fall through under its own weight. Bench marks should not be placed in such material unless you are sure that they are supported on permanently wet ground. If an additional load is placed on the ground near the bench mark, the surface may subside several feet under some conditions.

If the area is under water, stakes may be driven with the heads above water, as shown in Figure 80, and braced to keep the tops from being forced apart. A catwalk is built for the operator. Both the instrument support and the catwalk may be quite unstable constructions, provided they are entirely separate. At the time an observation is made, there

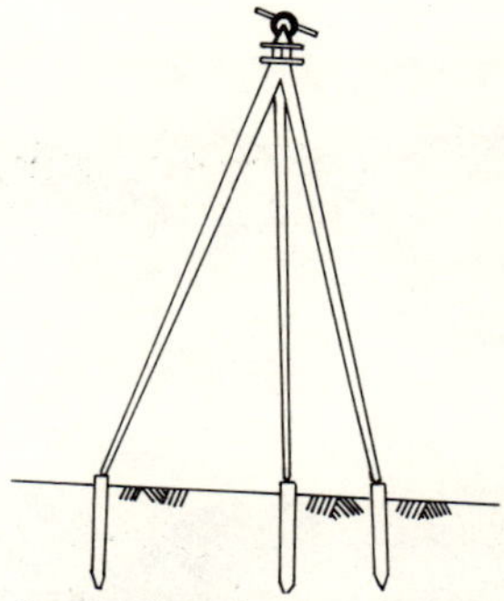

Figure 79. Soft ground setup.

should be no contact between the operator and the instrument, so that the catwalk may move freely under shifting load without affecting the instrument.

THE MACHETE

The machete is a long chopping knife, made in a number of patterns and sizes. The surveyor's machete, shown in Figure 81, has a 22-in. blade. For all but the heaviest cutting, such as trees over 2 in. in diameter, it is preferable to an axe or bush hook. The working chainman can carry a machete without greatly affecting his work; it is safer to carry the knife

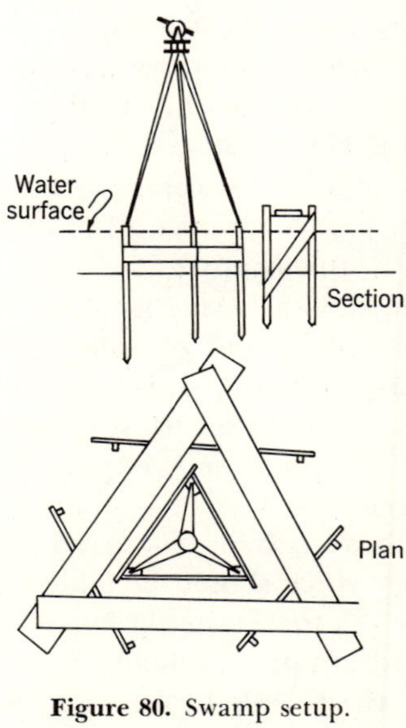

Figure 80. Swamp setup.

Figure 81. Surveyor's machete and scabbard. (Courtesy of the Collins Co.)

in a scabbard than in your belt (this also saves belts), but if the tool is frequently used, the scabbard can get in the way. Simple tools usually require greater skill of the operator than do complicated ones, and the machete is no exception. It is one of the most dangerous hand tools in use.

Any tool used for chopping is dangerous to the user, but the machete is capable of turning around and following the user after he has released it and has turned his back to run away! Slipping out of a sweaty hand, it can spear an unseen bystander 30 ft away, and a man can break his own leg with it.

The machete is the largest knife in common use; it is a chopping, not a cutting instrument. The Collins Company has manufactured this tool so long that in some parts of the world "Collins" is the name for a machete; the tools sold in the United States have Spanish-language labels, and they are now manufactured by this company in four Latin American plants. Machete may mean also a number of sizes and types of knives other than the one shown; in this country, the most common are shorter tools, such as the cane-knife and the Army machete. To check a machete for length, reach the floor with it and consider your ability to cut all day in that position. For most efficient use, brush should be cut very near the ground; if your machete cannot do this, it is too short.

The machete, like other chopping tools, requires frequent sharpening, and a file is a necessary accessory. Use of a stone is dangerous and unnecessary, as a very sharp edge is dangerous and quickly becomes dull anyway. Two safety features which should be added to the stock tool are illustrated in Figure 82. First, a thong should be attached, either through a hole in the handle, or tied around the blade next to the handle. This thong is passed around the wrist while in use and prevents your machete

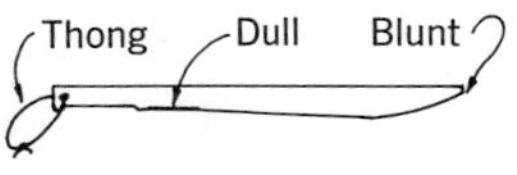

Figure 82. Safety features of the machete.

from leaving your hand. By using a slip knot, the machete can be carried on the wrist, leaving the hand free. When cutting overhead, the handle frequently slips when you hit a branch or vine near the handle, rotating the knife so that the edge returns directly to your face or hand. Completely out of control, propelled by its own rotation and by the elasticity of a branch, it may return point first to any part of your body. A machete flying in this manner can penetrate several inches into your body.

The point of the machete is rarely used, and should be clipped off as shown. This reduces damage if it should strike you point first. The first 6 in. of the blade next to the handle is practically unusable; if a branch is struck by this part, the machete will practically be torn from the hand. If this portion of the blade is unsharpened the danger of cutting one's self is reduced, and carrying the knife in the belt is less of a problem. The chainman is a skilled tradesman, and use of unskilled help is not only expensive in time consumed, but dangerous for the man and the rest of the crew. While a man is learning to use a machete, he should be watched as carefully as an army recruit on the rifle range.

PLUMBING WITH THE TRANSIT

If the standards, or side brackets, which support the telescope trunnions of a transit are at different heights, even very slightly, the instrument will point differently depending on the vertical angle of the telescope. This may be checked as shown in Figure 83. Set up near a tall building or flagpole, and sight at some point on the top which is easily identified; then point to the ground some distance away and set a nail in

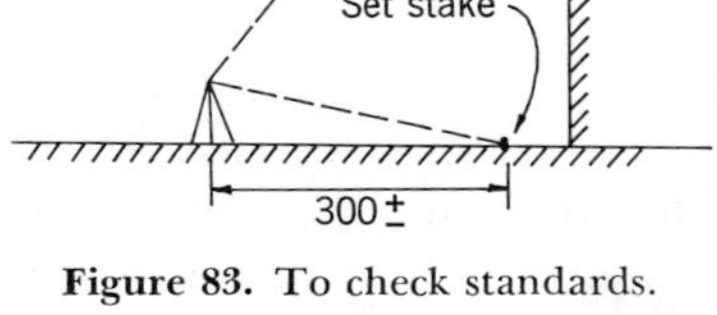

Figure 83. To check standards.

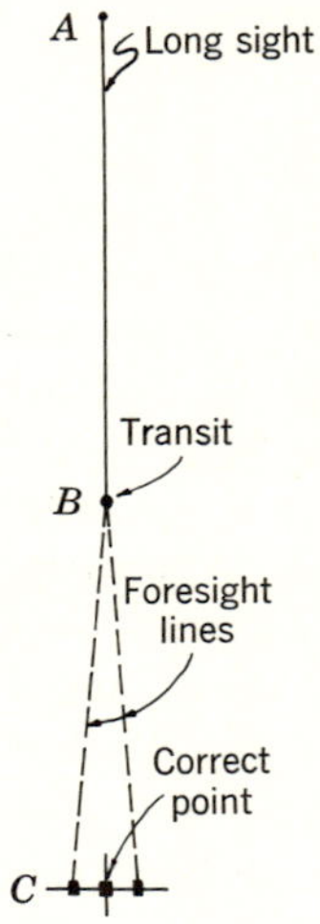

Figure 84. Double-centering.

a stake. This point should be directly in line with the higher point. Then turn the instrument 180°, plunge the telescope, and again sight on the high point, with the telescope now in the reversed position. Again sight on the stake set in the ground; if the instrument is in adjustment, the second point on the ground will be in the same place as the first. If there are two points, sight the instrument halfway between the two, and raise to look at the higher sight. Raise or lower the trunnion by the adjustment screw on the standard until the hair is on the upper point. Repeat for checking; it may be necessary to make several successively smaller adjustments.

DOUBLE CENTERING

If the line of sight of the telescope is not exactly at right angles with the trunnions that support it when it rotates, the instrument will point in a slightly different direction upon plunging. This is illustrated in Figure 84. The long sight is on a point as far away as can be found; plunge the instrument and set a point several hundred feet from the transit. The two points should be in line with the transit point. You could check this by moving up to the point you had set, and lining up the three points to see if they are really in line, but there is a quicker way. Turn the instrument 180° and sight again on the long sight; then plunge a second time. The telescope will now be in the direct position. Set a second point; if there is any error in the instrument, and there usually is, this second point will differ from the first. The true point is halfway between the two. If the transit is on a hill, so that the entire line cannot be seen through from one end, this procedure must be used to extend the line.

Since this sighting is usually done at a considerable distance the *double center* signal previously mentioned is used. The first point is set using the *give me line* signal repeatedly until the point is right; then the rod is revolved in the *double center* signal. The rodman usually can see the instrument man turn his back for the new backsight, or the instrument man may wave a flag in the *OK* signal. In any case, the rodman knows that if he immediately gets line again the instrument man missed his signal. To avoid the possibility that the instrument man has missed the *double center* signal, the rodman may continue this signal until he gets an *OK* signal from the instrument man. The rodman then calls for line, and sets the second point. If, on completion of the correct point, it is desired to turn an angle to it, or to set on it for other reasons, the rodman puts his rod on the true point (which would immediately bring a movement signal from the instrument man) and refuses to move as the instrument man directs; instead, the rodman passes his hand up and down the rod indicating the *this is line* signal. If the rod is to be set on line, the instrument man gives *OK,* and the rodman moves back, calls for line, and sets the rod on the true line. Alternately, if the point is not to be flagged, he may give a *move up* signal to the instrument man. This sequence of signals not only uses nearly all the commonly used visual signals, given at the maximum possible range, but it can be completely fouled up if one signal is missed. If the instrument man fails to see the *double center* signal, he will continue to give line, and the rodman may accept this as the second line, assuming the instrument to be in nearly perfect adjustment. Then, when he gives the instrument man a *move up* signal, the instrument man will refuse, and give the *double center* signal himself. The rodman should realize what has happened, replies with another *double center* signal, and proceeds to obtain the second line as before. Both people must know what is expected, and realize what the other has done.

Walkie-talkie sets can be used for such situations, but the signals mentioned are adequate, and the use of signals avoids carrying additional equipment.

Since the double centering error does not affect the accuracy of the instrument in other respects, and can be corrected for as above, it is very often tolerated. It may be corrected by a crosshair adjustment, but the student is advised to do this only under professional supervision. Older instruments have a spider web crosshair (from the finer wrappings of the spider's eggs) with a brass ring, which may be deformed; modern instruments have scribed marks on a glass plate. It is necessary to use the double centering procedure anyway, as a perfectly adjusted instrument may go out of adjustment at any time, and an error in extending the line may not be discovered by subsequent operations. Few instruments will remain in perfect adjustment for long periods.

WIGGLING OR JUGGLING IN

The preceding adjustment may be made by one man alone. He must keep the two points *A* and *C* in Figure 84 fixed, and change the point *B*; that is, move the instrument until he is in line with one position of the telescope, and set a point under the instrument. He then reverses the telescope and repeats the process. This operation is seldom used for adjustment, but is often used to set a line between two points, when a line of sight cannot be obtained from one to the other.

If *A* and *C* are already located, for example, but you cannot see from *A* to *C,* set up at *B* and find a point by trial and error that is on line. Usually wiggling in is done because *B* is on a hill or because *A* and *C* are too far apart for good sighting. *Wiggling* refers to the repeated trials that are necessary, making several setups and moving the instrument on the foot plate until the transit is exactly on line.

SHOOTING IN GRADES

Shooting in grades is the method of setting grades for a sloped line by inclining the telescope parallel to the sloped line. There are several variations of the method; the simplest application is shown in Figure 85. The tops of stakes *A* and *B* have been placed to a grade, or the same distance above or below a grade line, by conventional leveling methods, and you are to set a row of stakes such as *C* and *D,* with their tops on this same line. Set up over point *A,* and measure the distance from the top of stake *A* to the instrument, with a rod. Set the target on that reading, or merely use a stick and put a mark at that elevation. Take the same rod to stake *B,* and set the telescope on the target. Either a level or transit may be used; if a level, use the leveling head to incline it off level. You may then line up the rod anywhere along the line, such as *C* and *D,* raise or lower the rod so the target is on the horizontal crosshair, and the bottom of the rod will be at grade.

This same method may be used to give grade anywhere in an area, if the grade is a flat plane; that is, any line on it must be straight, although not necessarily level or at the same slope as any other line. On the finish grade map, contours are straight and the same distance apart over such an

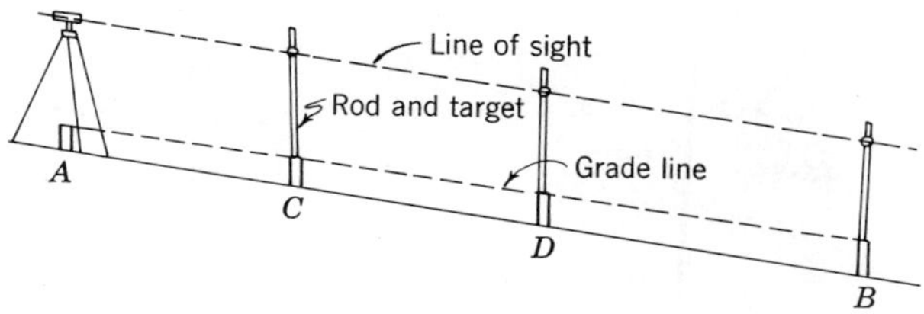

Figure 85. Shooting in grades.

area. The instrument is set by trial and error on two points in the area and it will be true for all other points.

A variation of this method is to set the telescope on the far point on a reading that will make the telescope parallel to the grade line, but to mark the stakes with cut and fill as outlined in Chapter 8. The calculation is simplified because the grade rod is the same for all stakes on a sloped line, rather than different for each point. Under both variations of the shot-in method, the stakes may be set at exact stations, may be paced out and set approximately on the stations, or may be set at random.

If the instrument is not set up directly over the line, it must be set to read on two points of the line; this may be done by repeated trials.

ELEVATIONS WITH THE CARPENTER'S LEVEL

For short distances, grade differences may be measured with a carpenter's level and a board, as shown in Figure 86. The differences between each pair of stakes is added to find the total difference. The stakes need not be set the same distance horizontally, so long as the board will reach them, and they need not be in a straight line.

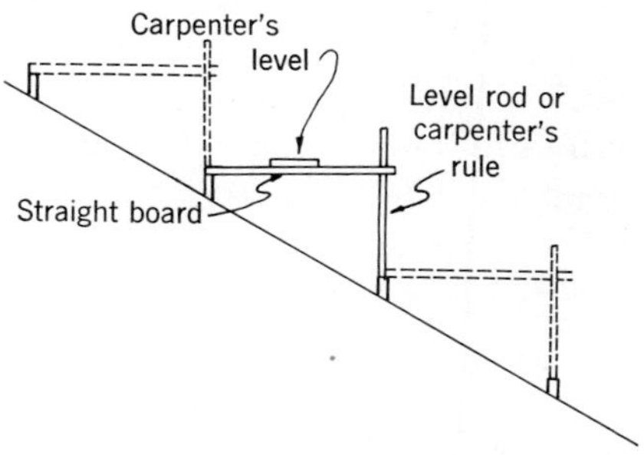

Figure 86. Elevation difference using carpenter's level.

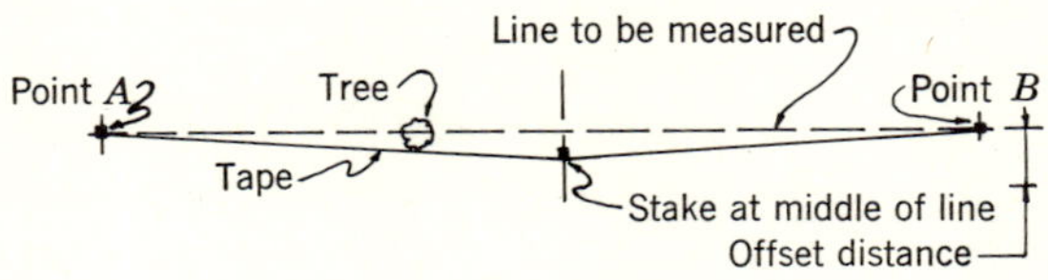

Figure 87. Measurement around tree.

MEASURING AROUND A TREE

The rule for computing error in a distance with one point off line may be used to find the true distance between two points when there is an obstruction between the points. Referring to Figure 87, the procedure is as follows.

1. Set a stake at the approximate middle of the line to be measured, as near the line as possible, but still in sight of both points, *A* and *B*.
2. Measure from *A* around the middle stake to *B*.
3. Line up your eye near point *B*, and measure the distance you must be from *B* to see *A*. This is the off-line distance.
4. Multiply this off-line distance by itself and by 50, and divide by the distance on the tape. This will give the correction, in hundredths, to be subtracted from the measured distance to get the true distance.

For example, if the offset distance is 3 ft and the distance measured on the tape is 60.00 ft, the correction is

$$\frac{3 \times 3 \times 50}{60} = \frac{450}{60} = 7.5$$

so the correction is rounded off to 0.08 ft and the distance is 59.92 ft. If the

$$L = \frac{40°00'}{5°00'} = 8, \qquad \text{or} \qquad 800.00 \text{ ft}$$

off-line distance is over 10% of the measured distance, the error in this method becomes significant. This is an application of the method of calculating error described in Chapter 2.

FIELD NOTES

Field note books are made with waterproof paper, and notes are kept with 6H (very hard) lead. This makes fine lettering possible without excessive sharpening of the pencil, and presses the figures into the paper so hard that the impressions may be read even after water has washed away the markings of a softer pencil. For sharpening the pencil, an emery board, used for women's finger nails and available in any drug store, is convenient.

There should be no erasures in notes of boundary surveys, because they greatly reduce or eliminate the usefulness of notes as evidence in court. Erasures in other books usually are not important, although some engineers object to any erasures.

ADJUSTING PINS

Surveying instruments are made with nuts to be turned with pins, and brass pins are available that will break before threads are stripped or the bolt twisted off. It is sometimes impossible to turn the nuts on old instruments with these pins, as threads become corroded and adjusting pins will break before the nuts will turn. Steel nails are too soft, bending before the nuts will turn. A dentist's pick, cut to size, will turn nearly any nut. If it is not used with care, you can shear off the bolts themselves in this way. This is a last resort, if a penetrating oil fails to loosen the nut.

FIELD EXERCISE NUMBER 15

This exercise is a review to check your ability to measure on a slope, and to satisfy yourself that the carpenter's level is accurate.

Select a sloping area and lay out an equilateral triangle large enough so there is about 7 ft elevation difference between two of the corners requiring you to break the chain for an accurate measurement. If the slope is steep enough, 60 ft is suitable for each side. In some areas a sloped area this high may not be available, in which case a window sill or other point can be used for one corner, even though a setup at that point is impractical.

Set two points of the triangle on the top of the slope at approximately the same elevation. Set up on one of these points, turn a 60° angle, and set the third point of the triangle toward the bottom of the slope. Measure the remaining side of the triangle, and set up at each of the other two points to check the angles. Error in angle should not be over 1′, and distance should not be off more than 0.01 ft. A triangle is used, because in a rectangle, such as used for the building layouts, errors due to improper plumbing tend to compensate each other. In a triangle, any difference in length between the side measured along the top of the slope, which should be between two points at approximately the same elevation, must be the same as the distance down the slope, or the angles will not be the same at each corner.

Next, determine the elevation difference between the top of one of the high stakes and the top of one of the low stakes using the transit and a level rod. Check this difference with a carpenter's level as described in this chapter. If equipment is available, this difference should also be determined with the Locke level and with the string level, to check the comparative accuracy of these tools. The carpenter's level method should be within 0.03 ft of the difference obtained by transit; usually it is much less. The Locke level reading should be off no more than 0.1 ft, and the string level by repeated trials and averaging, should be within 0.03 ft.

A few minutes should also be spent in checking the instrument for double-centering and accuracy in dropping the vertical from the top of a building. This entire exercise should be accomplished in about two hours by students who have completed previous work.

Chapter 15

THE CIRCULAR CURVE

Streets, highways, lot lines, and other alignments follow a circular curve; that is, a portion of a circle. The conventional designations for points of a circular curve are shown in Figure 88.

PI indicates the *point of intersection* of the straight lines, or *tangents,* which extend beyond the ends of the curve. Traverse computations of the survey, using straight lines, are made to this point.

The change in direction of the straight lines, or tangents, is variously called *delta* and marked Δ, the *central angle* marked *I,* or the *total deflection* of the tangents. Numerically, these are all the same, but may refer to measurements at different places. The angle between the *radii* (*R*) is the central angle, the angle between the tangents at the PI is the delta angle, and the total change in direction of the project centerline (along the curve) is the total deflection.

SETTING THE BEGINNING AND END OF CURVE

Suppose you are given the radius of curvature, *R* above, from design specifications or from the drawings of the project. This distance, for highways, is usually between 2,000 and 5,000 ft. You can measure the deflection angle, and should do so even though it is already on the plans. You must know the station of the PI, or can measure this from some other point on the tangent.

The PC, or *point of curvature,* is the point where the alinement changes from a straight line to a curve, and the PT, or *point of tangency,* is the point where it changes back from a curve to a straight line. These points are set by measuring along the tangents from the PI the *tangent distance* *T*. From the triangle above it may be seen that

$$T = R \tan \frac{\Delta}{2}$$

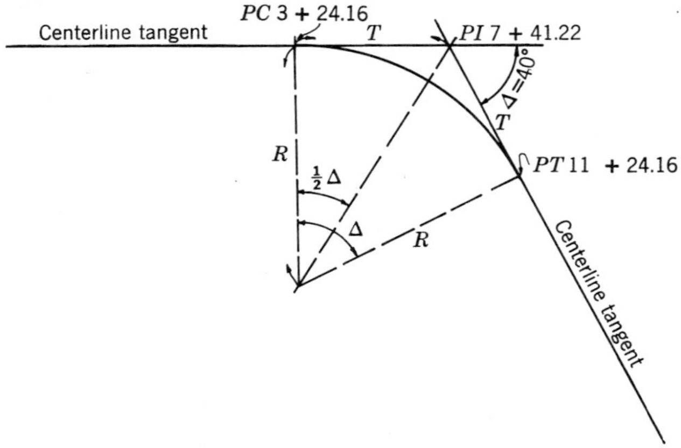

Figure 88. Simple curve.

Since six-place tables must be used for any accuracy in curve work, logarithmic computations are usually made when curves must be figured in the field. In this example, taking R as given to be 1,145.92 ft, and $\Delta = 40°00'$,

$$\begin{aligned} \log R \quad & = 3.05\,913 \\ +\log\tan\ \Delta/2 & = 9.56\,107 \\ \log T \quad & = \overline{2.62\,020} \\ T \quad & = 417.06 \text{ ft} \end{aligned}$$

The station designation of the PC is determined by subtracting this distance from the station of the PI, as

$$\begin{aligned} \text{PI} & = 7 + 41.22 \\ T & = 4 + 17.06 \\ \text{PC} & \quad \overline{3 + 24.16 \text{ ft}} \end{aligned}$$

The PT station cannot be determined in this way, but is figured by adding the distance around the curve to the station of the PC.

DEGREE OF CURVE

From the above information, the curve can be staked by swinging a tape from the center of curvature; this method is used for short-radius

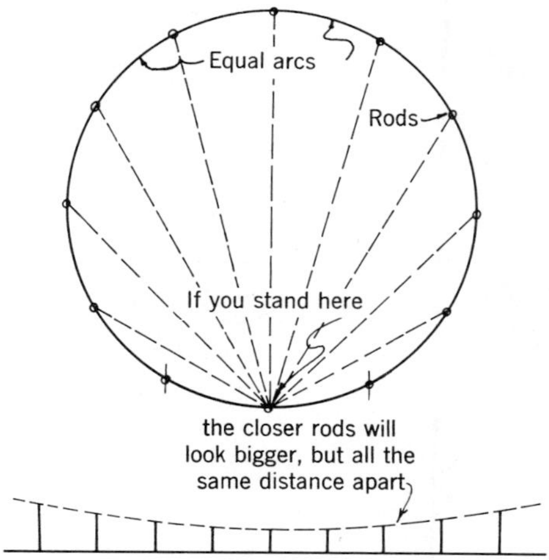

Figure 89. Arc angles from point on a curve.

curves such as curved buildings and street curbs. When the radius is so great that this method cannot be used, as for highway curves, the points along the curve are set by measuring from the PC or PT and using a different angle for each station. These angles, measured from a point on the curve as well as from the PC or PT are *station deflection angles.*

This method of layout is possible because of the geometric proposition that the angle subtended by an arc is, at the center of the circle, twice the angle subtended at a point on the circumference of a circle. The angle between the ends of any 100-ft arc will look the same as any other arc of the same length. If you are judging the distance away by the apparent distance between the ends, all the arcs will appear to be at the same distance, although the distance may actually vary from 100 ft to half a mile (Figure 89).

This illusion is caused by the fact that the nearer arcs are at an angle nearly in line with your line of sight, so that the ends appear close together (ends appearing close together caused the arc to appear more distant); and the arcs on the far side of the circle are across your line of sight, making the ends appear further apart (and the arc appear closer). If you are standing on a point on a 1000-ft circle that has a rod around it every 100 ft, the distance between any two adjacent rods will appear to be the same. If you are at the center of the circle, the angle of the ends of the arcs will still be the same, and will be just twice as large as if you were standing on the curve.

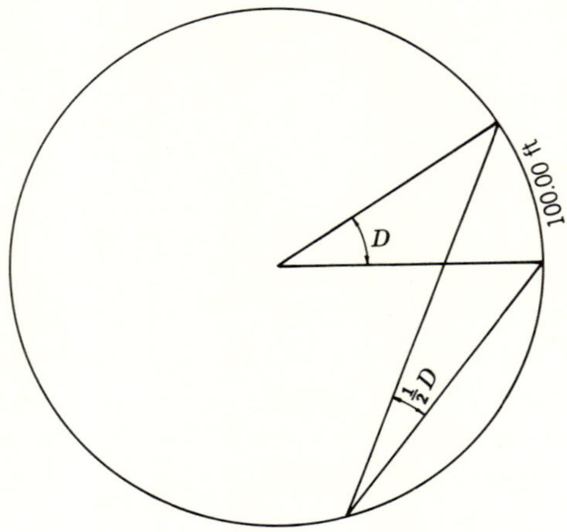

Figure 90

Using this relationship, a curve can be laid out from *any* point on the circle's circumference, using equal chords and equal deflections from any other point. Since the standard interval is a 100-ft station, the angle made at the center of the circle by a 100-ft arc is defined as the *degree of curvature, D*. The angle between the ends of a 100-ft arc, measured at a point on the curve, will always be ½ *D,* as in Figure 90.

This *D,* degree of curvature, is the deflection per 100-ft of arc along the curve, and this deflection between successive chords equals the central angle of that arc (Figure 91).

OTHER DEGREE OF CURVE DEFINITIONS

The surveyor must be careful when using curve tables that furnish information without detailed calculation, since there are definitions of degree of curve other than the one given above. In past years, degree of

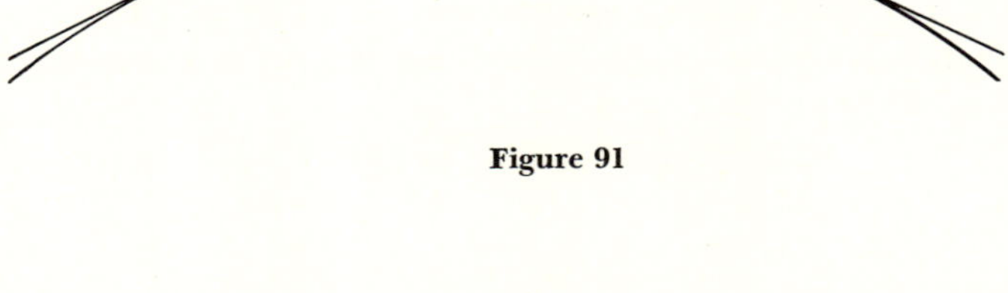

Figure 91

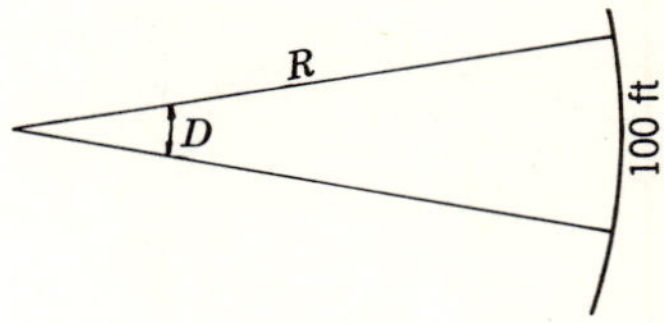

Figure 92

curve has been applied to the central angle of a 100-ft chord, an arc of 100.007 ft, or to a chord distance that varies with the degree of curve. If one follows the definition consistently, any definition will work, but be sure the tables you use are those for the kind of degree of curve you are using.

DEGREE OF CURVATURE AND RADIUS

The degree of curvature and radius are directly proportional to each other, which may be proven by reference to Figure 92, whereby

$$\text{circumference } C = 2\pi R$$

$$D = \frac{100}{C} \times 360°$$

and

$$D = \frac{100}{2\pi R} \times 360 = \frac{5729.58}{R}$$

The constant 5729.58, used for conversion of degree of curve to radius in the above formula, represents the radius, in ft, of a 1° curve. In the illustrated example

$$R = 1{,}145.92 \text{ ft}$$

$$D = \frac{5729.58}{1145.92} = 5°\ 00'$$

LENGTH OF CHORD

From Figure 91, it is evident that since the curve deflects D degrees per station, the total deflection is D times the length in stations, and the length of curve $L = \dfrac{\triangle}{D}$ in stations of 100 ft.

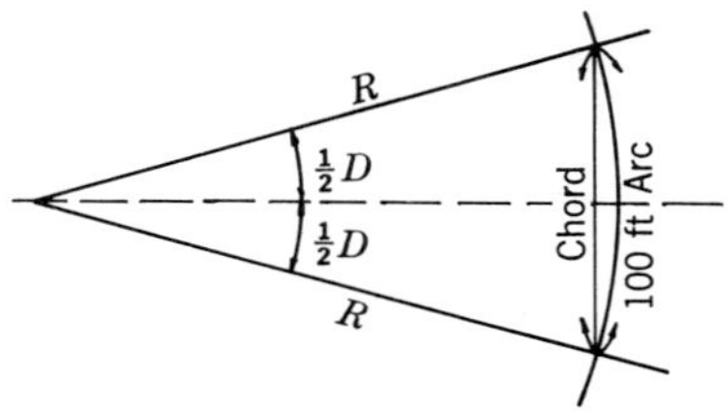

Figure 93

And for our example

$$L = \frac{40°00'}{5°00'} = 8, \text{ or } 800.00 \text{ ft}$$

Notice that since the length of curve is needed to at least five significant figures, D must be computed to thousandths of a degree.

The station of the PT is determined by adding the curve length (800 ft) to the station of the PC (3 + 24.16), giving 11 + 24.16.

Since you cannot measure along the arc, you must determine the chord distance corresponding to the arc distance between stakes, whether this be 100 ft, 50 ft, 25 ft or, as at the ends of the curve, an odd distance.

Assuming, from Figure 93, that the curve is to be staked at 100 ft intervals, the actual chord distance to be measured may be determined by

$$1/2 \text{ chord} = R \sin 1/2\ D$$

so

$$\text{chord} = 2R \sin 1/2\ D$$

Substituting values for a 100-ft chord

$$\begin{aligned} \log 2 &= 0.30103 \\ \log R &= 3.05913 \\ \log \sin 2°30' &= 8.63968 \\ \hline \log \text{chord} &= 1.99984 \\ \text{chord} &= 99.962 \text{ ft.} \end{aligned}$$

For arc lengths other than 100 ft, this expression becomes

$$1/2 \text{ chord} = 2R \sin\left(1/2\ D \times \frac{\text{arc length}}{100}\right)$$

This distance of 99.96 ft is the actual distance measured in a straight line (chord) to set the stations, which are 100 ft apart since the distance would be measured along the arc.

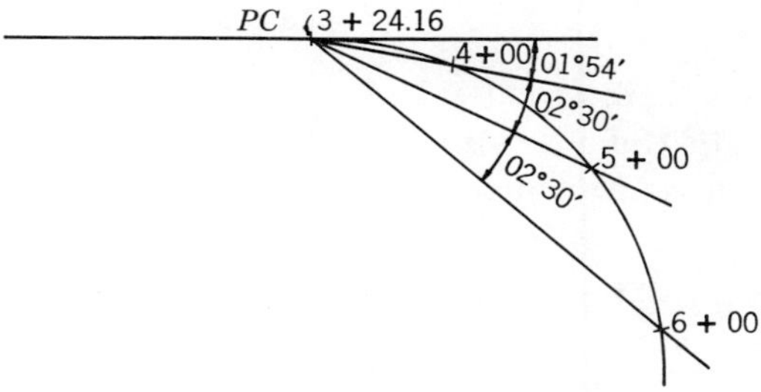

Figure 94

DEFLECTION ANGLES FOR LAYOUT

The first station to be set, 4 + 00, will be 75.84 ft from the PC. The correction for a 75.84-ft arc may be taken as 0.02′ without appreciable error, giving a chord for measurement of 75.82 ft, or the chord may be calculated. This correction is *not* proportional to the arc length.

The final step is to make a list of angles for each station, so that the curve may be staked without interruption for calculations.

Assuming we start from the PC and set 0° on the transit plates looking at the PI, we would turn $\frac{1}{2}$ D for each 100 ft of arc along the curve, as shown in Figure 94. However, the first distance is 75.84 ft of arc, not 100 ft, and the deflection is proportional. The deflection for station 4 + 00 is then $0.7584 \times \frac{1}{2} D = 0.7584 \times 2°30' = 01°54'$ (this may be a slide rule computation).

For each successive station, $\frac{1}{2}$ D is added to the first reading, rather than resetting on zero, so the angles and notes are as shown in Figure 94.

The last difference, for 24.16 ft, is $.2416 \times 2°30' = 36'$; adding this to 11 + 00 gives 20°00′, or half the total central angle. *This check must not be neglected.*

COMMON STUDENT ERRORS

The following errors are made most often by students.

1. Using some other arc length than 100 ft for computation of degree of curve, merely because stakes are being placed at some other interval. The result is to give a radius that depends not on the actual curve to be laid out, but on the stake interval. *The interval between stakes does not affect the computation of radius or of degree of curve.* The interval between stakes affects only the deflection difference for that interval.

2. Doubling the deflection angles, and using the deflection angle for

each station from the PC as the degree of curve, rather than half the degree as it should be.

3. Failing to check the total calculation by comparing the computed deflection angle for the PT with half the total deflection angle of the tangents or half the central angle.

LAYOUT FROM POINTS ON A CURVE

It may not be practical or convenient to run a curve from one setup of the transit. It may be run partially from each end, until the lines meet in the middle, or the transit may be set up on intermediate points on the curve. In any case, you use the same deflection angles as though you were set up on the PC, so no further calculations are necessary when another point is selected for the transit setup. Sometimes, as when a structure is on a curve, you may be required to accurately determine a line at right angles to the curve, which means it must be at right angles to the line tangent to the curve.

There are two basic rules to follow in this procedure.

1. *When sighting on any point on the curve, the plates on the transit must read the deflection angle for the point sighted on.*

2. *When sighting back on the line, the telescope must be in the reversed position, and when sighting forward, it must be in the direct position.*

If you clamp the lower motion as specified above, you will also be able to point along the tangent by setting on the plates the deflection angle for the transit location.

If the curve is to the right, all readings must be to the right, and vice-versa. This is true even though you may be set up on the PT and running the curve in backwards. The second rule above, that the instrument must be reversed when looking backward on the line, is not necessary if you do not use a backsight. That is, if you are always taking a sight in the same direction you are chaining, regardless of whether it is forward or backward on the line, it does not matter what the telescope position is; you set all angles, including the initial one, without plunging the instrument.

The rule above may be proven by plane geometry, and is illustrated, using the notes just completed, in Figure 95. Here, you have set stations 4, 5, 6, 7, and 8 from the PC. There is then an interference in the line of sight, so you set a tack point (usually curve stations are merely stakes) at station 7 and move forward to that station. Since the deflection angle is 0°00′ for the PC, you set this angle on the plates; and since the PC is behind you on the line and you intend to go forward, you plunge the tele-

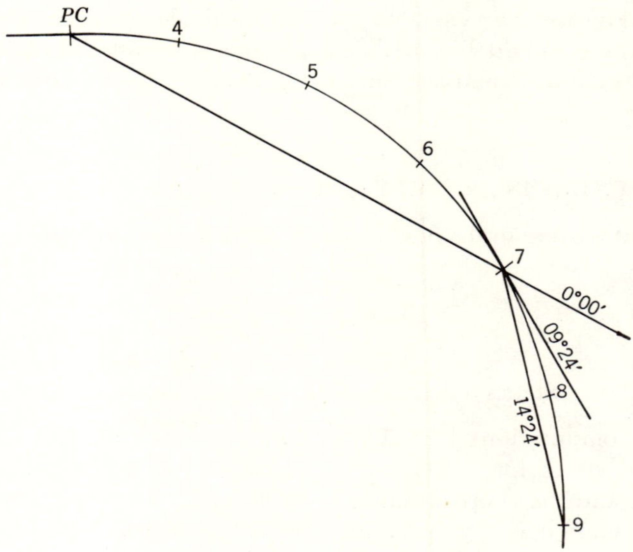

Figure 95. Setup on a curve.

scope to the reversed position. Now using the lower motion screw, so the reading is still 0 on the plates, you sight on the PC with 0° on the plates. You now have the lower motion oriented to the curve. To proceed with the curve, you plunge the telescope to the erect position (so it will be pointed in the direction you are going), and with the upper motion clamp and screw lay off the angle for 9, the next point to be set, at 14°24′. You then proceed around the balance of the curve, using the same angles as though you were set up at the PC.

If while at this station 7, you want the tangent to the curve, set the deflection for station 7 (09°24′) on the instrument. If you want a line at right angles to the curve at this point, set the angle at 99°24′ or 189°24′.

CLASSROOM EXERCISE NUMBER 6

Compute the radius, tangent distance, stations of PC, PT, chord length for 100-ft arc, and write deflection angles to lay out a curve, given that

Central angle = 07°00′
Degree of curve = 01°00′
PI station = 10 + 50.00

If this exercise is completed during one classroom period, and the deflection data for the next field exercise is calculated before the next field period, there should be little delay in the field work. If this exercise is done as homework, and not thoroughly understood, the field work will be difficult.

FIELD EXERCISE NUMBER 16

Lay out a curve in the field from the following information.

Central angle $= 25°00'$ Rt.
Degree of curve $= 10°00'$
PC station $= 0 + 00$

Stakes to be set at 50-ft intervals on the curve

Show computations for all data, including the tangent distance and length of chord for a 50-ft arc. The tangent distance, for checking, is 127.03 ft and the chord distance is 49.98 ft. If four-place tables are used, your calculation will vary slightly from these figures.

Notes with curve data and deflections should be prepared in advance; these deflections are from the PC, but you are to move up at an intermediate point on the curve for practice.

Procedure is as follows.

1. Set a stake and tack point at any convenient point, for the PI.
2. Measure the tangent distance, in any convenient direction, and set the PT.
3. Turn the central angle, measure off the other tangent distance, and set the PC.
4. Set up on PC and run in half of curve.
5. Move up on station near middle of curve, and run in remaining stations.
6. Report closing error, both for distance and angle, in the notes.

Tack-point all stations on the curve; as it is quite sharp, any errors in alignment will be increased as you go around the curve, and cause larger errors in your closure. You should be able to close this curve, on flat ground, within 0.03 ft.

This exercise requires three-man parties, and moving up in the middle of the curve provides an opportunity to change positions.

Chapter 16

LAND SURVEYING

TITLE

Title refers to a claim or right to use of property. A title to real estate must be registered in an appropriate office—usually the county courthouse. A person buying such a title is entitled to find on record all sales of the land and is not required to take physical possession in order to acquire the title.

The law of real estate property differs from other property law in important ways because of the inability to move the property and, therefore, to clearly identify it. One cannot pick up a piece of land and say "This is mine." There will always be an element of doubt as to the exact ownership. The attempt to remove such doubt as to who is the true owner and what are the claims by other persons is the work of the lawyer; the attempt to remove doubt as to the size and location of the property is the work of the land surveyor.

Title is not necessarily an attribute of land, and land does not necessarily exist because there is title. Titles are created by landowners who write a description of land and sell this description to others; land is created only by nature. Unfortunately these parties do not always work together.

An *abstract* is a history of the title of land. It originates with the person first able to hold possession against all comers, although even this person may have claimed it under nominal deed from another. The abstract is a typewritten copy of all papers on file that purports to show the legitimacy of the transfer of the title over past years. A person selling land usually furnishes an abstract; the buyer engages an attorney to examine this abstract and to give an opinion as to whether a title actually exists, and if the seller has a right to transfer the title.

Title insurance is a guarantee that the buyer has proper title, and is an agreement to compensate the buyer if the title should later prove to be faulty. Unlike other types of insurance, it is not risk insurance; that is, the insurer does not undertake to underwrite a risk. He may do so, but it is his intention to insure only titles that he has examined and of which he is fully satisfied that no risk of faulty title exists.

Title insurance is often an alternative to the seller's furnishing an abstract. In past years, sellers with unclear titles have sometimes set up their own title insurance firms in order to sell property. The buyer must, of course, satisfy himself that the insurer is a responsible firm.

SURVEY

The preceding discussion has no direct connection with land surveying, and is mentioned here only to emphasize this fact. Title insurance may give a person considerable peace of mind, but it does not necessarily give him a piece of land. Remember that title is a creation of man; an abstract by an attorney or the title insurance policy will carry a stipulation that any conditions "for which examination or survey of the property would be necessary to ascertain" are not covered. These conditions include the existence of the property.

If I write a deed for a lot located two miles out in the Atlantic Ocean, it could conceivably be a valid title, a valid deed, and may be an insurable title. If it happens to be a half-mile under water, that is unfortunate. If it happens that I have previously sold the same tract to a different person under another description, this is the buyer's tough luck. He may sue me for his payment, but he is not entitled to the property.

It is this fact that property may be sold twice, which gives rise to most boundary disputes; the land is inadvertently sold twice, under different descriptions. It is not the business of the surveyor to determine who is the owner of land, although this may result from his work; it is his business to retrace previous survey lines, or to define described lines. In either case, accuracy in measurement is the least important part of his work; a competent skilled workman can make measurements, but land surveying is a highly specialized profession.

If a survey is perfectly measured, the boundary is incorrect!

You should fully understand the terms used in this statement. *Perfect* is a theoretical quality; no person can make a perfect measurement. Consequently, a perfect measurement would not agree with any previous measurement. *A boundary* is a line which has been previously laid out in some way. For a boundary line to be correct, it must be in exactly the same place as it was when originally established.

You have, therefore, a contradiction in terms—a perfect line is unique, and there can be but one. A boundary is a repetition, and for it to be correct there must be two of them in the same place; the original and the resurvey. Therefore, the perfect line can never be the boundary. Since precision in measurement is an attempt at perfection, it may be an attempt to establish an incorrect boundary. If a line was run incorrectly, the surveyor attempts to retrace it; if it appears from the established part of a survey that the original surveyor used a tape that was 6 in. short in 100 ft, this is the tape-length that the following surveyor must use to retrace the line.

This is more than a play on words; the difference between the surveyor's work and the popular conception of the surveyor rests on these definitions. We can improve accuracy of our surveying methods, but we cannot force them on men long dead; for their boundaries, we must use their measurements.

• ORIGIN OF BOUNDARIES

As any person owning property may sell any part of it, and a landowner may sell any part of his holdings. His intent becomes the property description for succeeding generations.

Before about 1800, land in the United States was described in the way the seller desired, or by local surveys. That is, the description was made in a city or estate, without reference to an overall location beyond the immediate area.

Many of these descriptions defined land by reference to natural objects such as trees, creeks, land divides, and rocks. Existing buildings were used, and monuments were set using very rough measurements or none at all. Land was transferred by "the north 40 acres," or the "east half"; in many cases no markers were set to mark the proposed boundaries. Tracts are still transferred in this way, and new (and doubtful) boundaries are constantly being created.

About 1800, the federal government, in charge of surveys of all public lands (when nearly all the land west of the Alleghanies was public) began a survey which is still incomplete. Land is divided into townships of 36 square miles and into sections of one square mile. Along section lines monuments were set at half-mile intervals.

In both of the above systems of description, it is the purpose of the later surveyor to determine as best he can the original markers; or, lacking markers, to determine the original intent of the description. If measurements were initially made, they serve only as a guide to location of the original markers. If descriptions are contradictory, the surveyor must clearly define his assumptions and the description he used, as the language is properly defined by the court, not by surveyor. The surveyor must be familiar with the language, in order to defend his interpretation; but the fact that a secondary (not the original) surveyor has interpreted descriptions contrary to legal precedent or legal determination does not change the original boundary.

The monuments set at half-mile intervals are "half-section corners"; since they are original corners, they are correct if found, regardless of whether they were initially set correctly. The half-section corner in the middle of the section is set by extending lines between opposite original half-section lines on the boundaries of the section, and may be corrected at any time.

Many sections vary radically from the prescribed size; this happens both because of error and because of circumstances outside the control of the original surveyor. Surveys may have been started from points dozens of miles away; sections may end on bodies of water. The resulting error is thrown in the last quarter mile. Consequently, the purchase of a quarter section gives no assurance that the purchase is a fourth of the sectional area. The rules for determination of these areas are not complicated and are available in most standard works on surveying.

• SUBDIVISIONS

Purchasers of sections, particularly in present cities, subdivided their holdings into lots or other smaller divisions and sold them by descriptions that intended to match the manner of subdivision. In most cases, these maps (subdivision plats) were recorded in the country courthouse; if the description of the property given the buyer refers to a map, this map has to be filed for record in order to determine if a clear title exists. That is, you have no title (much less, any property) if you have a deed for "lot 1, McCarthy subdivision," when there is no McCarthy subdivision.

These subdivisions initially were sketches, filed with the deed to show the property being transferred. Many of them have no dimensions. It is the problem of the surveyor to locate the stakes, if these were originally set when the land was sold; if not, he must then arrive at some reasonable apportionment of the property. Unfortunately, there is often no record of whether stakes were set at time of sale; someone may discover such stakes years later, changing all boundaries set without reference to the original stakes. Later, the filing of plats was defined by statute, the form of plat was defined and a land surveyor was required to prepare them.

Land may still be subdivided by "metes and bounds" descriptions; that is, by describing the property in relation to existing markers. City lots of any size and location may be described in this way, and some developers have used such descriptions to avoid various legal obligations by the seller.

OCCUPATION

In order to prevent old survey lines from being constantly disputed, various statutes grant property rights to a person occupying the property. The requirements are highly technical, and summarized in the briefest form are listed here.

1. Occupation must be open and flagrant, and apparent to the public. Planting, mowing, fencing, constitute such occupation, but such use of property as grazing of cattle may not be sufficient. The intent is that the owner, if reasonably prudent, must be aware of the occupation.
2. Occupation must be "under color of title." That is, the person occupying the property must have a claim to the property; he may not merely occupy it. He must have some reason to believe he is the owner, and must act in good faith.
3. Occupation must be for a specified period of time. This varies from as short a period as seven years in Florida, to twenty in some states.

Public agencies may claim streets by occupation, and in this case it is not necessary to have a title claim. When no deeded right-of-way exists, road departments claim property to the adjoining fences or as far as they have mowed or built ditches. Owners may lose title to property by leaving it available for public use; some owners allow their property to be used by the public, but mark it with signs or close it for several days a year.

EVIDENCE OF LAND LINES

Lines of occupation, such as fences, are often used to establish the location of property lines—either section lines or other property lines. This use of occupation is entirely different from occupation for title of abutting landowners. A fence may be the best evidence of a section line and, therefore, it becomes a boundary; but a line may be proven to be erroneous after the statutory period of occupation has run. Such a line remains the boundary for abutting landowners, but has no influence in establishing the original land line that is used to establish other boundaries in the area.

REESTABLISHING SUBDIVISION LINES

Property lines within subdivision must be established at the original location, even though this may have been in error. If a landowner in error subdivides property that is not his, and establishes lots on his neighbor's property, the survey may not be corrected. The lot owners have bought lots according to the original survey; those lots which fall in the proper tract have a good title; others have a good title, but no property. It may be (and often happens) that the original landowner will keep part of the land he intended to sell, and sell land he did not own.

Subdivision plats often indicate how errors are to be distributed. It was common practice to mark ± on the last lot of each block; this saved computation and provided a place to distribute errors. The lot width was often scaled from the plat, and if there has been any error either in the survey or in plotting the subdivision plat the actual dimension of the lot may be far from the dimension indicated. If all lots are dimensioned, differences are distributed proportionately among the lots.

REGISTRATION

Civil engineers are responsible for topographical surveying and mapping; boundary surveying is done by land surveyors. Each may do work of the other when incidental to the practice of their own work. One person may register as both land surveyor and civil engineer, and pay a single annual fee, but there are separate examinations. These rules vary among the various states.

A person may qualify for land surveying registration by education, experience, and examination, according to state laws. The surveying examination includes both technical and legal matters well beyond the scope of this manual. Engineering survey courses include much of the material included in land surveying examinations, but few graduate engineers could pass the examination without both further study and experience. The necessary study is not difficult, and the books and typical examinations are readily available for persons who desire to enter the profession. In any case, a person should not attempt to do independent land surveying work without several years of experience in that type of surveying.

Both registered engineers and land surveyors may be granted permission by statute to go over the property of others when making surveys. Such entry is then not trespass, and they are not liable to arrest or civil action.

SUMMARY

The surveyor determines facts, not title. If he is given a land description that is a boundary, then he lays out a boundary; but he does not declare it to be a boundary. The surveyor retraces a description given him and does not attempt to determine title to the property. He may survey a boundary for adjacent landowners who have differing land descriptions, showing an overlap; so far as he is concerned, both surveys are correct. The determination of the correct description then becomes a matter of law, in which the surveyor would appear as an expert witness regarding the facts involved.

Measurements may or may not be correct on original plats, and if they are incorrect they must be repeated in the way they were originally made. Consequently, surveys have very little relationship with the immediate lot dimensions; the entire area must be considered.

In addition, it must be remembered that there are usually no public files showing surveyor's previous measurements or records; there is no assurance that any survey will not later be found in error by someone who has better information.

A surveyor, like any professional, makes no guarantee of his work beyond due care; and if bonded, such bond covers the surveyor, not the owner. The surveyor is bonded to be responsible for his legal liability, which is not a guarantee of boundaries he surveys.

FIELD EXERCISE NUMBER 17

This exercise is a final field examination, repeating early simple exercises under controlled conditions and with time recorded. The use of this test at the end of the course gives a basis of grading students on their ability to use the instrument, serves as a review of the early part of the course, and encourages additional practice with the instruments, when this is practical.

The exercise will taken an average of about 1 hour and 45 minutes; good students can complete it in an hour; and some slow ones may take two and a half hours. If a student takes longer than this, it indicates he has not learned the exercise initially. Because it is not only simple, but self-checking, there is no reason for wrong answers; and the basis for grading is largely either the time taken by each student, or his ability to remember how to do the work when he has a time limit. In either case, it is a test of both manual dexterity and of the student's probable ability to perform on the job.

It is suggested that four to five students take this test at a time, starting about fifty minutes apart. This requires as many transits and as many levels as there are students in each shift. The better students should start first, so that the later ones will not be delayed by over-running the earlier ones.

Assuming there are five students at a time, the instructor sets out five chains, five boxed transits, five boxed levels, and sets five points for transit setup. Level rods are set as they were for Field Exercise Numbers 4 and 6, in locations far enough apart that students on one exercise do not interfere with those on another.

Each student must do the following steps.

1. Unloop a chain and pick it up again, starting to let it out from the zero end, regardless of how it is rolled.
2. Unbox a transit, set up on a stake, and take the readings required by Field Exercise Number 6 (stadia readings on fixed rods). He records the number of his stake (one to five) for future checking by the instructor, boxes the instrument, and returns it to the place he found it.
3. Unbox a level, and read the level rods as in Field Exercise Number 4, but make one circuit rather than two.
4. Calculate the stadia readings recorded in step number 2. If there are errors he may return to the transit, if available, and check faulty readings. If the class is large, this check may not be practical. Field notes for this operation are kept as before.

The transit work is performed first, because otherwise one student would catch up with and interfere with another; the level work is much shorter. Calculations are delayed until after all the field work has been completed, so that another group of students may start on the same instruments.

RECOMMENDED READINGS

The working surveyor needs to continue his study well beyond the scope of this manual and what he will learn on the job, if he is to advance. The following current books are particularly recommended.

Brinker and Taylor, *Elementary Surveying* (International Textbook Co., Scranton, Pa., 1957). An excellent, complete, and well written text on engineering, including suggested exercises and field notes. There are but six pages on construction surveys, although the author notes that 60% of all surveying manhours are spent on location-type surveys giving line and grade. Because each construction project introduces individual problems, he limits textbook coverage of the topic. Circular curves, vertical surveys, and land surveys are included. 550 pp.

F. William Pafford, *Handbook of Survey Notekeeping* (John Wiley and Sons, New York, 1962). An excellent guide for notekeepers and party chiefs who want to know how notes are kept for a variety of conditions. It has a large number of illustrations of typical notes for different kinds of work. 140 pp.

Frank Clark, *A Treatise on the Law of Surveying and Boundaries* (Bobbs-Merrill Co., Indianapolis, Ind., 1939). The outstanding authority on surveying law, and a primary source for surveying practice and land surveyor examinations. Clearly written, but also contains ample legal citations. 894 pp.

William C. Wattles, *Land Survey Descriptions* (Title Insurance and Trust Company, Los Angeles, Calif., 1956). An excellent technical discussion of land descriptions, errors, and definitions of words used in land descriptions. 127 pp.

Curtis M. Brown, *Boundary Control and Legal Principles* (John Wiley and Sons, New York, 1957). A plainly written and quite detailed guide to a large number of specific problems encountered by the land surveyor. 275 pp.

There are many texts in engineering surveying; practically every publisher has at least one that he periodically revises. The best ones usually are the oldest, which have been updated, but still retain the older information and name. John Wiley and Sons, for example, publishes Breed and Hosmer's *Elementary Surveying,* which was first written in 1906. There are numerous out-of-print texts that are excellent, often with information about use and adjustment of instruments that is not included in newer books; some of these published in the late 1800's and early 1900's are of current value. The field engineer is advised to buy such books when available, usually in second-hand stores.

The American Surveyor and Photogrammetrist, a monthly magazine, publishes information of interest to the practicing surveyor as well as listing new books and other literature in the field. It is available from the Farrar Publishing Co., Suite 1028-30, National Press Building, Washington, D.C.

Field books, a variety of fillers including the double page fillers previously described, and a variety of data sheets for the field book are available from *Lefax Company, 9th and Samson Streets, Philadelphia, Pa.* This firm furnishes trigonometric, logarithmic, curve and stadia tables in the field book size.

Schools offering home study courses in surveying are listed in a free *Directory of Accredited Home Study Schools,* available from the National Home Study Council, 1601 18th Street, N. W., Washington, D.C. 20009.

Appendix

Four-Place Table of Logarithms

	0	1	2	3	4	5	6	7	8	9
10	0000	0043	0086	0128	0170	0212	0253	0294	0334	0374
11	0414	0453	0492	0531	0569	0607	0645	0682	0719	0755
12	0792	0828	0864	0899	0934	0969	1004	1038	1072	1106
13	1139	1173	1206	1239	1271	1303	1335	1367	1399	1430
14	1461	1492	1523	1553	1584	1614	1644	1673	1703	1732
15	1761	1790	1818	1847	1875	1903	1931	1959	1987	2014
16	2041	2068	2095	2122	2148	2175	2201	2227	2253	2279
17	2304	2330	2355	2380	2405	2430	2455	2480	2504	2529
18	2553	2577	2601	2625	2648	2672	2695	2718	2742	2765
19	2788	2810	2833	2856	2878	2900	2923	2945	2967	2989
20	3010	3032	3054	3075	3096	3118	3139	3160	3181	3201
21	3222	3243	3263	3284	3304	3324	3345	3365	3385	3404
22	3424	3444	3464	3483	3502	3522	3541	3560	3579	3598
23	3617	3636	3655	3674	3692	3711	3729	3747	3766	3784
24	3802	3820	3838	3856	3874	3892	3909	3927	3945	3962
25	3979	3997	4014	4031	4048	4065	4082	4099	4116	4133
26	4150	4166	4183	4200	4216	4232	4249	4265	4281	4298
27	4314	4330	4346	4362	4378	4393	4409	4425	4440	4456
28	4472	4487	4502	4518	4533	4548	4564	4579	4594	4609
29	4624	4639	4654	4669	4683	4698	4713	4728	4742	4757
30	4771	4786	4800	4814	4829	4843	4857	4871	4886	4900
31	4914	4928	4942	4955	4969	4983	4997	5011	5024	5038
32	5051	5065	5079	5092	5105	5119	5132	5145	5159	5172
33	5185	5198	5211	5224	5237	5250	5263	5276	5289	5302
34	5315	5328	5340	5353	5366	5378	5391	5403	5416	5428
35	5441	5453	5465	5478	5490	5502	5514	5527	5539	5551
36	5563	5575	5587	5599	5611	5623	5635	5647	5658	5670
37	5682	5694	5705	5717	5729	5740	5752	5763	5775	5786
38	5798	5809	5821	5832	5843	5855	5866	5877	5888	5899
39	5911	5922	5933	5944	5955	5966	5977	5988	5999	6010
40	6021	6031	6042	6053	6064	6075	6085	6096	6107	6117
41	6128	6138	6149	6160	6170	6180	6191	6201	6212	6222
42	6232	6243	6253	6263	6274	6284	6294	6304	6314	6325
43	6335	6345	6355	6365	6375	6385	6395	6405	6415	6425
44	6435	6444	6454	6464	6474	6484	6493	6503	6513	6522
45	6532	6542	6551	6561	6571	6580	6590	6599	6609	6618
46	6628	6637	6646	6656	6665	6675	6684	6693	6702	6712
47	6721	6730	6739	6749	6758	6767	6776	6785	6794	6803
48	6812	6821	6830	6839	6848	6857	6866	6875	6884	6893
49	6902	6911	6920	6928	6937	6946	6955	6964	6972	6981
50	6990	6998	7007	7016	7024	7033	7042	7050	7059	7067
51	7076	7084	7093	7101	7110	7118	7126	7135	7143	7152
52	7160	7168	7177	7185	7193	7202	7210	7218	7226	7235
53	7243	7251	7259	7267	7275	7284	7292	7300	7308	7316
54	7324	7332	7340	7348	7356	7364	7372	7380	7388	7396

(*continued*)

	0	1	2	3	4	5	6	7	8	9
55	7404	7412	7419	7427	7435	7443	7451	7459	7466	7474
56	7482	7490	7497	7505	7513	7520	7528	7536	7543	7551
57	7559	7566	7574	7582	7589	7597	7604	7612	7619	7627
58	7634	7642	7649	7657	7664	7672	7679	7686	7694	7701
59	7709	7716	7723	7731	7738	7745	7752	7760	7767	7774
60	7782	7789	7796	7803	7810	7818	7825	7832	7839	7846
61	7853	7860	7868	7875	7882	7889	7896	7903	7910	7917
62	7924	7931	7938	7945	7952	7959	7966	7973	7980	7987
63	7993	8000	8007	8014	8021	8028	8035	8041	8048	8055
64	8062	8069	8075	8082	8089	8096	8102	8109	8116	8122
65	8129	8136	8142	8149	8156	8162	8169	8176	8182	8189
66	8195	8202	8209	8215	8222	8228	8235	8241	8248	8254
67	8261	8267	8274	8280	8287	8293	8299	8306	8312	8319
68	8325	8331	8338	8344	8351	8357	8363	8370	8376	8382
69	8388	8395	8401	8407	8414	8420	8426	8432	8439	8445
70	8451	8457	8463	8470	8476	8482	8488	8494	8500	8506
71	8513	8519	8525	8531	8537	8543	8549	8555	8561	8567
72	8573	8579	8585	8591	8597	8603	8609	8615	8621	8627
73	8633	8639	8645	8651	8657	8663	8669	8675	8681	8686
74	8692	8698	8704	8710	8716	8722	8727	8733	8739	8745
75	8751	8756	8762	8768	8774	8779	8785	8791	8797	8802
76	8808	8814	8820	8825	8831	8837	8842	8848	8854	8859
77	8865	8871	8876	8882	8887	8893	8899	8904	8910	8915
78	8921	8927	8932	8938	8943	8949	8954	8960	8965	8971
79	8976	8982	8987	8993	8998	9004	9009	9015	9020	9025
80	9031	9036	9042	9047	9053	9058	9063	9069	9074	9079
81	9085	9090	9096	9101	9106	9112	9117	9122	9128	9133
82	9138	9143	9149	9154	9159	9165	9170	9175	9180	9186
83	9191	9196	9201	9206	9212	9217	9222	9227	9232	9238
84	9243	9248	9253	9258	9263	9269	9274	9279	9284	9289
85	9294	9299	9304	9309	9315	9320	9325	9330	9335	9340
86	9345	9350	9355	9360	9365	9370	9375	9380	9385	9390
87	9395	9400	9405	9410	9415	9420	9425	9430	9435	9440
88	9445	9450	9455	9460	9465	9469	9474	9479	9484	9489
89	9494	9499	9504	9509	9513	9518	9523	9528	9533	9538
90	9542	9547	9552	9557	9562	9566	9571	9576	9581	9586
91	9590	9595	9600	9605	9609	9614	9619	9624	9628	9633
92	9638	9643	9647	9652	9657	9661	9666	9671	9675	9680
93	9685	9689	9694	9699	9703	9708	9713	9717	9722	9727
94	9731	9736	9741	9745	9750	9754	9759	9763	9768	9773
95	9777	9782	9786	9791	9795	9800	9805	9809	9814	9818
96	9823	9827	9832	9836	9841	9845	9850	9854	9859	9863
97	9868	9872	9877	9881	9886	9890	9894	9899	9903	9908
98	9912	9917	9921	9926	9930	9934	9939	9943	9948	9952
99	9956	9961	9965	9969	9974	9978	9983	9987	9991	9996

Appendix

Trigonometric Functions

Degrees	Sin	Cos	Tan	Degrees	Sin	Cos	Tan
0	0.0000	1.0000	0.0000	45	0.7071	0.7071	1.0000
1	0.0175	0.9998	0.0175	46	0.7193	0.6947	1.0355
2	0.0349	0.9994	0.0349	47	0.7314	0.6820	1.0724
3	0.0523	0.9986	0.0524	48	0.7431	0.6691	1.1106
4	0.0698	0.9976	0.0699	49	0.7547	0.6561	1.1504
5	0.0872	0.9962	0.0875	50	0.7660	0.6428	1.1918
6	0.1045	0.9945	0.1051	51	0.7771	0.6293	1.2349
7	0.1219	0.9925	0.1228	52	0.7880	0.6157	1.2799
8	0.1392	0.9903	0.1405	53	0.7986	0.6018	1.3270
9	0.1564	0.9877	0.1584	54	0.8090	0.5878	1.3764
10	0.1736	0.9848	0.1763	55	0.8192	0.5736	1.4281
11	0.1908	0.9816	0.1944	56	0.8290	0.5592	1.4826
12	0.2079	0.9781	0.2126	57	0.8387	0.5446	1.5399
13	0.2250	0.9744	0.2309	58	0.8480	0.5299	1.6003
14	0.2419	0.9703	0.2493	59	0.8572	0.5150	1.6643
15	0.2588	0.9659	0.2679	60	0.8660	0.5000	1.7321
16	0.2756	0.9613	0.2867	61	0.8746	0.4848	1.8040
17	0.2924	0.9563	0.3057	62	0.8829	0.4695	1.8807
18	0.3090	0.9511	0.3249	63	0.8910	0.4540	1.9626
19	0.3256	0.9455	0.3443	64	0.8988	0.4384	2.0503
20	0.3420	0.9397	0.3640	65	0.9063	0.4226	2.1445
21	0.3584	0.9336	0.3839	66	0.9135	0.4067	2.2460
22	0.3746	0.9272	0.4040	67	0.9205	0.3907	2.3559
23	0.3907	0.9205	0.4245	68	0.9272	0.3746	2.4751
24	0.4067	0.9135	0.4452	69	0.9336	0.3584	2.6051
25	0.4226	0.9063	0.4663	70	0.9397	0.3420	2.7475
26	0.4384	0.8988	0.4877	71	0.9455	0.3256	2.9042
27	0.4540	0.8910	0.5095	72	0.9511	0.3090	3.0777
28	0.4695	0.8829	0.5317	73	0.9563	0.2924	3.2709
29	0.4848	0.8746	0.5543	74	0.9613	0.2756	3.4874
30	0.5000	0.8660	0.5774	75	0.9659	0.2588	3.7321
31	0.5150	0.8572	0.6009	76	0.9703	0.2419	4.0108
32	0.5299	0.8480	0.6249	77	0.9744	0.2250	4.3315
33	0.5446	0.8387	0.6494	78	0.9781	0.2079	4.7046
34	0.5592	0.8290	0.6745	79	0.9816	0.1908	5.1446
35	0.5736	0.8192	0.7002	80	0.9498	0.1736	5.6713
36	0.5878	0.8090	0.7265	81	0.9877	0.1564	6.3138
37	0.6018	0.7986	0.7536	82	0.9903	0.1392	7.1154
38	0.6157	0.7880	0.7813	83	0.9925	0.1219	8.1443
39	0.6293	0.7771	0.8098	84	0.9945	0.1045	9.5144
40	0.6428	0.7660	0.8391	85	0.9962	0.0872	11.430
41	0.6561	0.7547	0.8693	86	0.9976	0.0698	14.301
42	0.6691	0.7431	0.9004	87	0.9986	0.0523	19.081
43	0.6820	0.7314	0.9325	88	0.9994	0.0349	28.636
44	0.6947	0.7193	0.9657	89	0.9998	0.0175	57.290
45	0.7071	0.7071	1.0000	90	1.0000	0.0000	——

INDEX